FIX-POINTS AND FACTORIZATION OF MEROMORPHIC FUNCTIONS

Chi-Tai Chuang
Peking University
China

Chung-Chun Yang
Naval Research Laboratory
USA

World Scientific
Singapore • New Jersey • London • Hong Kong

Published by

World Scientific Publishing Co. Pte. Ltd.,
P O Box 128, Farrer Road, Singapore 9128
USA office: 687 Hartwell Street, Teaneck, NJ 07666
UK office: 73 Lynton Mead, Totteridge, London N20 8DH

ISBN 981-02-0008-0

Printed in Singapore by JBW Printers & Binders Pte. Ltd.

FIX-POINTS AND FACTORIZATION OF MEROMORPHIC FUNCTIONS

PREFACE

This book is essentially the English translation of the authors' Chinese mathematical monograph which was published by Peking University Press in 1988. The study of the theory of meromorphic functions has had a history of more than fifty years in China, since 1930's. However, until very recently, most of the research efforts were devoted to the investigations of intrinsic properties of the functions. Little attention has been paid to the applications of the theory of meromorphic functions to other types of mathematical problems.

The fix-points and factorization in the composite sense of meromorphic functions are two closely related applied topics of the theory of meromorphic functions. Julia and Fatou were the pioneers in the study of the fix-points of iterates of functions by using classical theory of normal families. Many of their results were refined and extended by contemporary complex analysts such as Baker, Hayman, and Chuang. The study of factorization theory of meromorphic functions gained momentum in late 1960's, notably due to the efforts of Gross, Yang, Ozawa, and Goldstein. Later, further progress was made by the joint effort of Urabe and Prokopovich in 1970's, and of Song, Noda, and Steinmetz in 1980's. Up to the present, Nevanlinna's theory of meromorphic functions and Wiman-Valiron type theorems remain the most powerful tools in attacking the problems raised in the two topics mentioned above, and further advances are awaited.

Numerous papers and two English reference books have been published on the two subjects: "Factorization of meromorphic functions" by Gross in

1972 and "Factorization theory of meromorphic function" edited by Yang in 1983. The material contained in these two books about fix-points and factorization aspects now appear to be out of date. Moreover, since then, many interesting results in these two subjects have been obtained, and new techniques have been developed. Also the work in these areas is done over the world now, particular in U.S.A., China, Japan, Germany, and U.S.S.R. The Chinese version of the book has thus been written for updating the research progress in these two subjects, and for graduate students majoring in complex analysis, and researchers who are interested in the value-distribution theory of meromorphic functions and its applications. The Chinese book has been used in seminar courses for graduate students at Peking University and studied by many Chinese mathematicians with great enthusiasm. We hope the English version will be welcome by more readers throughout the world and more fruitful research accomplishments in these two subjects and their related ones will result.

The book contains four chapters and an appendix. The first two chapters reintroduce fundamentals of Nevanlinna's theory of meromorphic functions and Montel's theory of normal families of holomorphic functions. The last two chapters introduce factorization theory and the relationship between the fix-points and factorization of a function; many recent results in factorization theory are reported and related open questions are raised for further studies. The appendix consists of some useful facts about the growth properties of a composite meromorphic function, properties on differential polynomials, and a simpler proof of an important result of Steinmetz on certain type of functional equations.

We would like to express our sincere thanks to the publisher, Dr. K.K. Phua for his endorsement of the project, Ms. H.M. Ho for her many helpful and instructive assistance and Joyce Tay for her excellent typing skill which transformed the handwritten manuscript into a pleasant looking book.

Chi-tai Chuang and Chung-chun Yang

CONTENTS

FIX-POINTS AND FACTORIZATION OF MEROMORPHIC FUNCTIONS

1

NEVANLINNA'S THEORY OF
MEROMORPHIC FUNCTIONS

1.1. INTRODUCTION

Since Nevanlinna published his fundamental paper on the theory of meromorphic functions in 1925, many subsequent works have been done by various authors, which, as a whole, constitute the modern theory of meromorphic functions. However in this book we are mainly interested in its applications to the study of fix-points and factorization of meromorphic functions. In this chapter, we first give a systematic sketch of the initial part of that theory, which consists of the definition and properties of the characteristic function, the first fundamental theorem and the second fundamental theorem as well as their applications to value-distribution of meromorphic functions. Next we give a complete proof of a classical theorem of Borel on systems of entire functions satisfying an identity. This theorem of Borel, which is often applied to the study of fix points and factorization of meromorphic functions, is deduced from a theorem of Nevanlinna on systems of meromorphic functions.

For further information on Nevanlinna's theory and its subsequent development, the reader is referred to the following books:

Nevanlinna, R., *Le Théorème de Picard-Borel et la Théorie des Fonctions Méromorphes*, Paris, 1929.

Hayman, W.K., *Meromorphic Functions*, Oxford, 1964.

Chuang, Chi-tai, *Singular Directions of Meromorphic Functions* (in Chinese), Beijing, 1982.

In this book, by meromorphic functions we always mean functions which are meromorphic in the complex plan (except at places explicitly stated). A meromorphic function is said to be transcendental, if it is not a rational function. An entire function may be considered as a meromorphic function which does not take the value ∞.

1.2. POISSON-JENSEN FORMULA

Theorem 1.1. Let $f(z)$ be a meromorphic function in a domain $|z| < R$ $(0 < R \leq \infty)$ non identically equal to zero. Consider a disk $|z| < \rho$ $(0 < \rho < R)$ and the zeros $a_\lambda (\lambda = 1, 2, \ldots, h)$ and the poles $b_\mu (\mu = 1, 2, \ldots, k)$ of $f(z)$ in the disk $|z| < \rho$, where each zero or pole appears as many times as its order. Then in the disk $|z| < \rho$ the following formula holds:

$$
\begin{aligned}
\log |f(z)| = \frac{1}{2\pi} \int_0^{2\pi} & \log |f(\rho e^{i\varphi})| \mathrm{Re} \left(\frac{\rho e^{i\varphi} + z}{\rho e^{i\varphi} - z} \right) d\varphi \\
& - \sum_{\lambda=1}^{h} \log \left| \frac{\rho^2 - \bar{a}_\lambda z}{\rho(z - a_\lambda)} \right| + \sum_{\mu=1}^{k} \log \left| \frac{\rho^2 - \bar{b}_\mu z}{\rho(z - b_\mu)} \right| .
\end{aligned}
\tag{1.1}
$$

Nevanlinna calls this formula the Poisson-Jensen formula.

Proof. To prove this formula we distinguish two cases:

$1° f(z)$ has no zero and pole on the circle $|z| = \rho$. As we know, if z_0 is a point of the disk $|z| < \rho$, then the function

$$
w = \frac{\rho(z - z_0)}{\rho^2 - \bar{z}_0 z}
$$

transforms the circle $|z| = \rho$ to the circle $|w| = 1$ and the interior of the circle $|z| = \rho$ to the interior of the circle $|w| = 1$. Consider the products

$$
P(z) = \prod_{\lambda=1}^{h} \frac{\rho(z - a_\lambda)}{\rho^2 - \bar{a}_\lambda z} , \quad Q(z) = \prod_{\mu=1}^{k} \frac{\rho(z - b_\mu)}{\rho^2 - \bar{b}_\mu z}
$$

and the function

$$
F(z) = f(z) \frac{Q(z)}{P(z)} .
\tag{1.2}
$$

$F(z)$ is a meromorphic function in the domain $|z| < R$, without zero and pole in the disk $|z| < \rho$, and such that

$$
|F(z)| = |f(z)|
\tag{1.3}
$$

on the circle $|z| = \rho$. R_1 ($\rho < R_1 < R$) being a number such that for $\rho < |z| < R_1$, $F(z)$ has no zero and pole, then $F(z)$ has no zero and pole in the disk $|z| < R_1$, and consequently $\log |F(z)|$ is a harmonic function in the disk $|z| < R_1$. By Poisson formula in the disk $|z| < \rho$, we have

$$\log |F(z)| = \frac{1}{2\pi} \int_0^{2\pi} \log |F(\rho e^{i\varphi})| \mathrm{Re}\left(\frac{\rho e^{i\varphi} + z}{\rho e^{i\varphi} - z}\right) d\varphi . \qquad (1.4)$$

In view of (1.3), the integral in (1.4) is equal to that in (1.1). Hence by (1.2), we get (1.1) from (1.4).

2° On the circle $|z| = \rho$, $f(z)$ has zero or pole. In this case, the integral in (1.1) is an improper integral. So we must first show that it has a meaning. Since

$$\left| \mathrm{Re}\left(\frac{\rho e^{i\varphi} + z}{\rho e^{i\varphi} - z}\right) \right| \leq \frac{\rho + |z|}{\rho - |z|} ,$$

it is sufficient to show this for the integral

$$\int_0^{2\pi} |\log |f(\rho e^{i\varphi})|| d\varphi . \qquad (1.5)$$

Consider a zero or pole $z_0 = \rho e^{i\varphi_0}$ on the circle $|z| = \rho$. In a small disk $|z - z_0| < \delta$, we have

$$f(z) = (z - z_0)^s g(z) ,$$

where s is an integer (positive or negative) and $g(z)$ is a function holomorphic and without zero in the disk $|z - z_0| < \delta$. It follows that for $\varphi_0 - \eta \leq \varphi \leq \varphi_0 + \eta (\eta > 0$ sufficiently small), we have

$$f(\rho e^{i\varphi}) = (\rho e^{i\varphi} - \rho e^{i\varphi_0})^s g(\rho e^{i\varphi}) ,$$
$$\log |f(\rho e^{i\varphi})| = s \log \rho + \log |e^{i\varphi} - e^{i\varphi_0}| + \log |g(\rho e^{i\varphi})| ,$$
$$|\log |f(\rho e^{i\varphi})|| \leq A + \log \frac{1}{|e^{i\varphi} - e^{i\varphi_0}|} ,$$

where A is a constant. Next making use of the relation

$$\lim_{\varphi \to \varphi_0} \left| \frac{e^{i\varphi} - e^{i\varphi_0}}{\varphi - \varphi_0} \right| = 1 ,$$

we have

$$|e^{i\varphi} - e^{i\varphi_0}| > \frac{1}{2}|\varphi - \varphi_0|,$$

$$|\log|f(\rho e^{i\varphi})\|| < A + \log 2 + \log \frac{1}{|\varphi - \varphi_0|}$$

for $0 < |\varphi - \varphi_0| \le \eta_1$ $(0 < \eta_1 < \eta)$. Since the integral

$$\int_{\varphi_0 - \eta_1}^{\varphi_0 + \eta_1} \log \frac{1}{|\varphi - \varphi_0|} d\varphi$$

has a meaning, this is also true for the integral

$$\int_{\varphi_0 - \eta_1}^{\varphi_0 + \eta_1} |\log|f(\rho e^{i\varphi})\|| d\varphi$$

and then for the integral (1.5).

Now let us show that (1.1) holds in the disk $|z| < \rho$. Consider a point z of this disk, which is not a zero or pole of $f(z)$. Let $\rho'(0 < \rho' < \rho)$ be a number such that the point z and $a_\lambda(\lambda = 1, 2, \ldots, h), b_\mu(\mu = 1, 2, \ldots, k)$ are all in the disk $|z| < \rho'$. Then, by case 1°, we have

$$\log|f(z)| = \frac{1}{2\pi} \int_0^{2\pi} \log|f(\rho' e^{i\varphi})| \operatorname{Re}\left(\frac{\rho' e^{i\varphi} + z}{\rho' e^{i\varphi} - z}\right) d\varphi$$

$$- \sum_{\lambda=1}^{h} \log\left|\frac{\rho'^2 - \bar{a}_\lambda z}{\rho'(z - a_\lambda)}\right| + \sum_{\mu=1}^{k} \log\left|\frac{\rho'^2 - b_\mu z}{\rho'(z - b_\mu)}\right|.$$

As $\rho' \to \rho$, the first and the second of the two sums in this formula tend respectively to the corresponding sums in (1.1). It can also be shown that the integral in this formula tends to the integral in (1.1). Hence (1.1) holds.

Assuming $f(0) \neq 0, \infty$ and setting $z = 0$ in (1.1), we get Jensen formula

$$\log|f(0)| = \frac{1}{2\pi} \int_0^{2\pi} \log|f(\rho e^{i\varphi})| d\varphi - \sum_{\lambda=1}^{h} \log \frac{\rho}{|a_\lambda|} + \sum_{\mu=1}^{k} \log \frac{\rho}{|b_\mu|}. \quad (1.6)$$

Denoting respectively by $n(r, \frac{1}{f})$ and $n(r, f)$ the number of the zeros and the number of the poles in the disk $|z| \le r$ $(0 < r < R)$ of $f(z)$ (each zero or pole being counted as many times as its order), we know that

$$\sum_{\lambda=1}^{h} \log \frac{\rho}{|a_\lambda|} = \int_{r_0}^{\rho} \left(\log \frac{\rho}{t}\right) dn\left(t, \frac{1}{f}\right),$$

where r_0 is a sufficiently small positive number. Integrating by parts, we see that the above integral is equal to

$$\left[\left(\log\frac{\rho}{t}\right)n\left(t,\frac{1}{f}\right)\right]_{r_0}^{\rho} + \int_{r_0}^{\rho}\frac{n\left(t,\frac{1}{f}\right)}{t}dt = \int_{r_0}^{\rho}\frac{n\left(t,\frac{1}{f}\right)}{t}dt = \int_0^{\rho}\frac{n\left(t,\frac{1}{f}\right)}{t}dt .$$

So we have

$$\sum_{\lambda=1}^{h}\log\frac{\rho}{|a_\lambda|} = \int_0^{\rho}\frac{n\left(t,\frac{1}{f}\right)}{t}dt$$

and similarly

$$\sum_{\mu=1}^{k}\log\frac{\rho}{|b_\mu|} = \int_0^{\rho}\frac{n(t,f)}{t}dt .$$

On the other hand, defining $\log^+ x$ $(x \geq 0)$ by

$$\log^+ x = 0 \ (0 \leq x < 1), \quad \log^+ x = \log x \ (x \geq 1) ,$$

we have

$$\log|f(z)| = \log^+|f(z)| - \log^+\left|\frac{1}{f(z)}\right| ,$$

$$\frac{1}{2\pi}\int_0^{2\pi}\log|f(\rho e^{i\varphi})|d\varphi = \frac{1}{2\pi}\int_0^{2\pi}\log^+|f(\rho e^{i\varphi})|d\varphi$$
$$- \frac{1}{2\pi}\int_0^{2\pi}\log^+\left|\frac{1}{f(\rho e^{i\varphi})}\right|d\varphi . \qquad (1.7)$$

(1.6) can therefore be written in the form

$$\frac{1}{2\pi}\int_0^{2\pi}\log^+|f(\rho e^{i\varphi})|d\varphi + \int_0^{\rho}\frac{n(t,f)}{t}dt$$
$$= \frac{1}{2\pi}\int_0^{2\pi}\log^+\left|\frac{1}{f(\rho e^{i\varphi})}\right|d\varphi + \int_0^{\rho}\frac{n\left(t,\frac{1}{f}\right)}{t}dt + \log|f(0)| . \qquad (1.8)$$

In the above Jensen formula, we have assumed that $f(0) \neq 0, \infty$. If the point $z = 0$ is a zero or pole of $f(z)$, we introduce the auxiliary function

$$f_1(z) = z^{-s}f(z) ,$$

where s is a positive or negative integer occurring in the development

$$f(z) = c_s z^s + c_{s+1} z^{s+1} + \dots \quad (c_s \neq 0) \tag{1.9}$$

in the neighborhood of the point $z = 0$. $f_1(z)$ is also meromorphic in the domain $|z| < R$ with $f_1(0) = c_s$. Noting that $n\left(r, \frac{1}{f_1}\right) = n\left(r, \frac{1}{f}\right) - n\left(0, \frac{1}{f}\right), n(r, f_1) = n(r, f) - n(0, f)$,

$$\frac{1}{2\pi} \int_0^{2\pi} \log |f_1(\rho e^{i\varphi})| d\varphi = \frac{1}{2\pi} \int_0^{2\pi} \log |f(\rho e^{i\varphi})| d\varphi - s \log \rho \tag{1.10}$$

$$s = n\left(0, \frac{1}{f}\right) - n(0, f) ,$$

and applying (1.8) to $f_1(z)$, we get, by making use of the relations (1.10), the formula

$$\frac{1}{2\pi} \int_0^{2\pi} \log^+ |f(\rho e^{i\varphi})| d\varphi + \int_0^\rho \frac{n(t, f) - n(0, f)}{t} dt + n(0, f) \log \rho$$

$$= \frac{1}{2\pi} \int_0^{2\pi} \log^+ \left|\frac{1}{f(\rho e^{i\varphi})}\right| d\varphi + \int_0^\rho \frac{n\left(t, \frac{1}{f}\right) - n\left(0, \frac{1}{f}\right)}{t} dt$$

$$+ n\left(0, \frac{1}{f}\right) \log \rho + \log |c_s| . \tag{1.11}$$

This formula which contains (1.8) as a particular case, is called Jensen-Nevanlinna formula.

1.3. CHARACTERISTIC FUNCTION

Let $f(z)$ be a meromorphic function in a domain $|z| < R$ $(0 < R \leq \infty)$ non identically equal to zero, and consider the formula (1.11). If in this formula we set $\rho = r$ and define

$$m(r, f) = \frac{1}{2\pi} \int_0^{2\pi} \log^+ |f(re^{i\varphi})| d\varphi , \tag{1.12}$$

$$N(r, f) = \int_0^r \frac{n(t, f) - n(0, f)}{t} dt + n(0, f) \log r , \tag{1.13}$$

then (1.11) may be written as

$$m(r, f) + N(r, f) = m\left(r, \frac{1}{f}\right) + N\left(r, \frac{1}{f}\right) + \log |c_s| \, . \tag{1.14}$$

Next we define

$$T(r, f) = m(r, f) + N(r, f) \quad (0 < r < R) \, . \tag{1.15}$$

Then (1.14) becomes

$$T(r, f) = T\left(r, \frac{1}{f}\right) + \log |c_s| \, . \tag{1.16}$$

Nevanlinna calls $T(r, f)$ the characteristic function of the function $f(z)$. In the particular case that $f(z)$ is identically equal to zero, $T(r, f) = 0$ is also defined, but (1.16) no longer holds.

Now we are going to prove a formula for $T(r, f)$ and deduce some of its properties. We start from the formula

$$\frac{1}{2\pi} \int_0^{2\pi} \log |\alpha - e^{i\theta}| d\theta = \log^+ |\alpha| \, . \tag{1.17}$$

This formula holds for any complex number α. In fact, this is evident, if $\alpha = 0$. If $\alpha \neq 0$, we apply (1.6) to the function $\varphi(z) = \alpha - z$ with $\rho = 1$ and get

$$\log |\alpha| = \frac{1}{2\pi} \int_0^{2\pi} \log |\varphi(e^{i\theta})| d\theta \quad (|\alpha| \geq 1) \, ,$$

$$\log |\alpha| = \frac{1}{2\pi} \int_0^{2\pi} \log |\varphi(e^{i\theta})| d\theta - \log \frac{1}{|\alpha|} \quad (|\alpha| < 1) \, .$$

In order to obtain the desired formula for $T(r, f)$, we assume that $f(z)$ is non-constant and $f(0)$ is finite. θ being a real number such that $e^{i\theta} \neq f(0)$, application of (1.7) and (1.8) to the function $f(z) - e^{i\theta}$ yields

$$\frac{1}{2\pi} \int_0^{2\pi} \log |f(re^{i\varphi}) - e^{i\theta}| d\varphi = N\left(r, \frac{1}{f - e^{i\theta}}\right) - N(r, f) + \log |f(0) - e^{i\theta}| \, ,$$

where $0 < r < R$. Next keeping r fixed and integrating both sides with respect to θ, we get

$$\frac{1}{2\pi} \int_0^{2\pi} d\theta \frac{1}{2\pi} \int_0^{2\pi} \log |f(re^{i\varphi}) - e^{i\theta}| d\varphi$$

$$= \frac{1}{2\pi} \int_0^{2\pi} N\left(r, \frac{1}{f - e^{i\theta}}\right) d\theta - N(r, f) + \frac{1}{2\pi} \int_0^{2\pi} \log |f(0) - e^{i\theta}| d\theta \ . \tag{1.18}$$

Then interchanging the order of integration of the integral on the left of (1.18) and making use of (1.17), we get

$$\frac{1}{2\pi} \int_0^{2\pi} d\theta \frac{1}{2\pi} \int_0^{2\pi} \log |f(re^{i\varphi}) - e^{i\theta}| d\varphi$$

$$= \frac{1}{2\pi} \int_0^{2\pi} d\varphi \frac{1}{2\pi} \int_0^{2\pi} \log |f(re^{i\varphi}) - e^{i\theta}| d\theta$$

$$= \frac{1}{2\pi} \int_0^{2\pi} \log^+ |f(re^{i\varphi})| d\varphi = m(r, f) \ ,$$

$$\frac{1}{2\pi} \int_0^{2\pi} \log |f(0) - e^{i\theta}| d\theta = \log^+ |f(0)| \ .$$

Finally from these relations and (1.18) we get the desired formula

$$T(r, f) = \frac{1}{2\pi} \int_0^{2\pi} N\left(r, \frac{1}{f - e^{i\theta}}\right) d\theta + \log^+ |f(0)| \ . \tag{1.19}$$

This formula is due to Cartan H.

In the proof of (1.19) we have assumed that $f(0)$ is finite. If the point $z = 0$ is a pole of $f(z)$, then by (1.7) and (1.11) we have

$$\frac{1}{2\pi} \int_0^{2\pi} \log |f(re^{i\varphi}) - e^{i\theta}| d\varphi = N\left(r, \frac{1}{f - e^{i\theta}}\right) - N(r, f) + \log |c_s| \ ,$$

where c_s is the coefficient in the development (1.9), and we get in the same way the formula

$$T(r, f) = \frac{1}{2\pi} \int_0^{2\pi} N\left(r, \frac{1}{f - e^{i\theta}}\right) d\theta + \log |c_s| \ . \tag{1.20}$$

From (1.19) or (1.20) we see easily that $T(r, f)$ is a non-decreasing function of r and a convex function of $\log r$. In fact, it is sufficient to show that this is true for $N(r, f)$. That $N(r, f)$ is a non-decreasing function of r is obvious. To see that it is a convex function of $\log r$, consider three values $r_j (j = 1, 2, 3)$ such that $0 < r_1 < r_2 < r_3 < R$. We have

$$N(r_2, f) - N(r_1, f) = \int_{r_1}^{r_2} \frac{n(t, f) - n(0, f)}{t} dt + n(0, f) \log \frac{r_2}{r_1}$$

$$= \int_{r_1}^{r_2} \frac{n(t, f)}{t} dt \leq n(r_2, f) \int_{r_1}^{r_2} \frac{dt}{t} = n(r_2, f) \log \frac{r_2}{r_1}$$

and similarly

$$N(r_3, f) - N(r_2, f) \geq n(r_2, f) \log \frac{r_3}{r_2} .$$

Consequently

$$\frac{N(r_2, f) - N(r_1, f)}{\log \frac{r_2}{r_1}} \leq \frac{N(r_3, f) - N(r_2, f)}{\log \frac{r_3}{r_2}} .$$

This implies

$$N(r_2, f) \leq \frac{\log \frac{r_3}{r_2}}{\log \frac{r_3}{r_1}} N(r_1, f) + \frac{\log \frac{r_2}{r_1}}{\log \frac{r_3}{r_1}} N(r_3, f) . \tag{1.21}$$

which shows that $N(r, f)$ is a convex function of $\log r$.

Now we are going to prove some inequalities for $m(r, f), N(r, f)$ and $T(r, f)$, which are often used. First of all, it is easy to see that the inequalities

$$\log^+ (\alpha_1 \alpha_2 \ldots \alpha_p) \leq \sum_{j=1}^{p} \log^+ \alpha_j , \tag{1.22}$$

$$\log^+ (\alpha_1 + \alpha_2 + \ldots + \alpha_p) \leq \sum_{j=1}^{p} \log^+ \alpha_j + \log p \tag{1.23}$$

hold for arbitrary $\alpha_j \geq 0$ $(j = 1, 2, \ldots, p)$. In fact, to see (1.22), it is sufficient to note that the left member of (1.22) is 0, if $\alpha_1 \alpha_2 \ldots \alpha_p < 1$ and is

$$\log(\alpha_1 \alpha_2 \ldots \alpha_p) = \sum_{j=1}^{p} \log \alpha_j \leq \sum_{j=1}^{p} \log^+ \alpha_j ,$$

if $\alpha_1 \alpha_2 \ldots \alpha_p \geq 1$. On the other hand, to see (1.23), let $\alpha = \max(\alpha_1, \alpha_2, \ldots, \alpha_p)$ and note that the left member of (1.23) does not exceed

$$\log^+ p\alpha \leq \log^+ \alpha + \log p .$$

Let $f_j(z)$ $(j = 1, 2, \ldots, p)$ be p meromorphic functions in a domain $|z| < R$ $(0 < R \leq \infty)$. From (1.22) and (1.23), we deduce immediately the inequalities

$$m(r, f_1 f_2 \ldots f_p) \leq \sum_{j=1}^{p} m(r, f_j) , \qquad (1.24)$$

$$m(r, f_1 + f_2 + \ldots + f_p) \leq \sum_{j=1}^{p} m(r, f_j) + \log p \qquad (1.25)$$

which hold for $0 < r < R$. On the other hand, if $f_j(0) \neq \infty$ $(j = 1, 2, \ldots, p)$, then we also have the inequalities

$$N(r, f_1 f_2 \ldots f_p) \leq \sum_{j=1}^{p} N(r, f_j) , \qquad (1.26)$$

$$N(r, f_1 + f_2 + \ldots + f_p) \leq \sum_{j=1}^{p} N(r, f_j) , \qquad (1.27)$$

$0 < r < R$. We give a proof of these inequalities only for the case $p = 2$, because then they are true in general by induction. Evidently it is sufficient to show that

$$n(t, f_1 f_2) \leq n(t, f_1) + n(t, f_2) , \qquad (1.28)$$

$$n(t, f_1 + f_2) \leq n(t, f_1) + n(t, f_2) , \qquad (1.29)$$

for $0 < t < R$. To see (1.28), consider a disk $|z| \leq t$ and first assume that $f_1(z)f_2(z)$ has no pole in this disk, then (1.28) is evident. Next assume that $f_1(z)f_2(z)$ has poles in this disk and let z_k $(k = 1, 2, \ldots, q)$ be all the distinct poles in this disk of $f_1(z)$ and $f_2(z)$. For each k, define $S_k^{(1)}$ as follows: $S_k^{(1)}$ is equal to the order of z_k, if z_k is a pole of $f_1(z)$; otherwise $S_k^{(1)} = 0$. Similarly define $S_k^{(2)}$ and $S_k^{(12)}$ with respect to $f_2(z)$ and $f_1(z)f_2(z)$. We see easily that

$$S_k^{(12)} \leq S_k^{(1)} + S_k^{(2)} \quad (k = 1, 2, \ldots, q)$$

and hence

$$\sum_{k=1}^{q} S_k^{(12)} \leq \sum_{k=1}^{q} S_k^{(1)} + \sum_{k=1}^{q} S_k^{(2)} .$$

This proves (1.28), because the three sums are respectively equal to the three terms in (1.28). (1.29) is proved in a similar manner.

It can also be shown that, if $R > 1$, then (1.26) and (1.27) hold for $1 \leq r < R$ without the assumption $f_j(0) \neq \infty$ $(j = 1, 2, \ldots, p)$.

Finally from (1.24)-(1.27), we get the inequalities

$$T(r, f_1 f_2 \ldots f_p) \leq \sum_{j=1}^{p} T(r, f_j) , \tag{1.30}$$

$$T(r, f_1 + f_2 + \ldots + f_p) \leq \sum_{j=1}^{p} T(r, f_j) + \log p \tag{1.31}$$

which hold under the same conditions as for (1.26) and (1.27).

For a function $f(z)$ holomorphic in a domain $|z| < R$ $(0 < R \leq \infty)$, the functions $T(r, f)$ and

$$M(r, f) = \max_{|z|=r} |f(z)| \tag{1.32}$$

are both important. It is interesting that they satisfy the inequality

$$T(r, f) \leq \log^+ M(r, f) \leq \frac{\rho + r}{\rho - r} T(\rho, f) \tag{1.33}$$

for $0 < r < \rho < R$. The first part of (1.33) is obvious, because

$$N(r, f) = 0, \quad T(r, f) = m(r, f) .$$

To prove the second part, we make use of (1.1) which now becomes

$$\log|f(z)| = \frac{1}{2\pi} \int_0^{2\pi} \log|f(\rho e^{i\varphi})|\mathrm{Re}\left(\frac{\rho e^{i\varphi} + z}{\rho e^{i\varphi} - z}\right) d\varphi$$
$$- \sum_{\lambda=1}^{h} \log\left|\frac{\rho^2 - \bar{a}_\lambda z}{\rho(z - a_\lambda)}\right| .$$

In this formula, setting $z = z_0$ where z_0 is a point of the circle $|z| = r$ such that $|f(z_0)| = M(r, f)$ and noting

$$0 < \mathrm{Re}\left(\frac{\rho e^{i\varphi} + z_0}{\rho e^{i\varphi} - z_0}\right) \leq \frac{\rho + r}{\rho - r} ,$$

we get

$$\log M(r, f) \leq \frac{1}{2\pi} \int_0^{2\pi} \log^+ |f(\rho e^{i\varphi})|\mathrm{Re}\left(\frac{\rho e^{i\varphi} + z_0}{\rho e^{i\varphi} - z_0}\right) d\varphi$$
$$\leq \frac{\rho + r}{\rho - r} T(\rho, f) .$$

1.4. FIRST FUNDAMENTAL THEOREM

Let $f(z)$ be a non-constant meromorphic function in a domain $|z| < R$ $(0 < R \leq \infty)$. Consider a finite value a and the development of $f(z) - a$ in the neighborhood of the point $z = 0$

$$f(z) - a = c_s z^s + c_{s+1} z^{s+1} + \ldots \quad (c_s \neq 0) .$$

By (1.16),

$$T(r, f - a) = T\left(r, \frac{1}{f - a}\right) + \log|c_s| \quad (0 < r < R) . \tag{1.34}$$

Let us compare $T(r, f)$ and

$$T(r, f - a) = m(r, f - a) + N(r, f - a) .$$

We have

$$N(r, f - a) = N(r, f)$$

and

$$m(r, f - a) \leq m(r, f) + \log^+ |a| + \log 2 \,,$$
$$m(r, f) \leq m(r, f - a) + \log^+ |a| + \log 2 \,.$$

Consequently

$$|m(r, f - a) - m(r, f)| \leq \log^+ |a| + \log 2$$

and

$$|T(r, f - a) - T(r, f)| \leq \log^+ |a| + \log 2 \,. \tag{1.35}$$

(1.34) and (1.35) yield

$$T\left(r, \frac{1}{f - a}\right) = T(r, f) + h(r) \,, \tag{1.36}$$

where

$$|h(r)| \leq |\log |c_s|| + \log^+ |a| + \log 2 \,.$$

Nevanlinna calls (1.36) the first fundamental theorem.

Furthermore, consider a function of the form

$$F(z) = \frac{af(z) + b}{cf(z) + d} \,,$$

where a, b, c, d are constants such that $ad - bc \neq 0$. If we regard F as a fractional linear function of f, it can be decomposed into several functions of the following forms

$$f_1(z) = f(z) + \beta, \quad f_2(z) = \alpha f(z), \quad f_3(z) = \frac{1}{f(z)} \,.$$

where β and $\alpha \neq 0$ are constants. Since the characteristic function of each of these functions differs from $T(r, f)$ by a bounded function, we get the following result: The difference $T(r, F) - T(r, f)$ is a bounded function. This is a complement of the first fundamental theorem.

Finally we make an investigation of the growth of $T(r, f)$, when $f(z)$ is a meromorphic function in the complex plane. Such a function $f(z)$ is briefly called a meromorphic function, as already mentioned at the beginning of this chapter.

We first prove that if $f(z)$ is a non-constant meromorphic function, then

$$\lim_{r \to \infty} T(r, f) = \infty . \tag{1.37}$$

In fact, if $f(0) = \infty$, then by the definition (1.13) of $N(r, f)$, evidently

$$\lim_{r \to \infty} N(r, f) = \infty$$

and *a fortiori* (1.37) holds. If $a = f(0) \neq \infty$, then

$$\lim_{r \to \infty} T\left(r, \frac{1}{f - a}\right) = \infty .$$

So (1.37) again holds by (1.36).

Next we are going to prove that if $f(z)$ is a transcendental meromorphic function, then

$$\lim_{r \to \infty} \frac{T(r, f)}{\log r} = \infty . \tag{1.38}$$

Distinguish two cases:

1° $f(z)$ has no pole. In this case, $f(z)$ is a transcendental entire function. In its development

$$f(z) = \sum_{n=0}^{\infty} a_n z^n$$

there is an infinite number of coefficients different from zero. It follows then from Cauchy inequalities

$$|a_n| r^n \leq M(r, f) \quad (r > 0, n = 0, 1, 2, \ldots) ,$$

that

$$\lim_{r \to \infty} \frac{M(r, f)}{r^p} = \infty$$

for each positive integer p. This implies

$$\lim_{r \to \infty} \frac{\log M(r, f)}{\log r} = \infty . \tag{1.39}$$

On the other hand, taking $\rho = 2r$ in (1.33), we have

$$\log^+ M(r, f) \leq 3T(2r, f) .$$

Hence (1.38) holds.

$2°$ $f(z)$ has poles. First assume that $f(z)$ has an infinite number of poles. Then from

$$N(r^2, f) \geq N(r^2, f) - N(r, f) \geq n(r, f) \log r \quad (r > 1) ,$$

we have

$$\lim_{r \to \infty} \frac{N(r, f)}{\log r} = \infty$$

and, *a fortiori*, (1.38). Next assume that $f(z)$ has only a finite number of poles $b_j (j = 1, 2, \ldots, k)$ whose orders are respectively $m_j (j = 1, 2, \ldots, k)$. Set

$$P(z) = \prod_{j=1}^{k} (z - b_j)^{m_j} , \quad g(z) = P(z) f(z) .$$

Then, remembering that $f(z)$ is not a rational function, $g(z)$ is a transcendental entire function, and hence (1.38) holds for $g(z)$. On the other hand, by (1.30),

$$T(r, g) \leq T(r, P) + T(r, f) \leq m \log r + K + T(r, f) \quad (r \geq 1) ,$$

where $m = \sum_{j=1}^{k} m_j$ and $K > 0$ is a constant. Hence (1.38) also holds.

Concerning the growth of a meromorphic function, an important notion is that of its order. The order ρ of a non-constant meromorphic function $f(z)$ is defined by

$$\rho = \overline{\lim_{r \to \infty}} \frac{\log T(r, f)}{\log r} . \tag{1.40}$$

We have $0 \leq \rho \leq \infty$, and we shall denote it by $\rho(f)$. When ρ is finite, then, for each positive number ε, on one hand there is a value r_0 such that

$$T(r, f) < r^{\rho + \varepsilon} \quad \text{for} \quad r > r_0 ,$$

and on the other hand there is a sequence of values r_n $(n = 1, 2, \ldots)$ tending to ∞ such that

$$T(r, f) > r^{\rho - \varepsilon} \quad \text{for} \quad r = r_n \ (n = 1, 2, \ldots) .$$

By a complement of the first fundamental theorem given above, for a function of the form

$$F(z) = \frac{a f(z) + b}{c f(z) + d} ,$$

where a, b, c, d are constants such that $ad - bc \neq 0$, the difference $T(r, F) - T(r, f)$ is a bounded function. Consequently the functions $F(z)$ and $f(z)$ have the same order.

Now consider the particular case that $f(z)$ is a non-constant entire function. Then by (1.33),

$$T(r, f) \leq \log^+ M(r, f) \leq 3T(2r, f)$$

and therefore

$$\varlimsup_{r \to \infty} \frac{\log T(r, f)}{\log r} = \varlimsup_{r \to \infty} \frac{\log \log M(r, f)}{\log r} .$$

This shows that, in the particular case of entire functions, the present definition of order is compatible with the classical one.

For a meromorphic function $f(z)$ which is a constant, we define its order $\rho(f) = 0$.

1.5. LOGARITHMIC DERIVATIVE

Let us return to Poisson-Jensen formula (1.1) and consider a point z_0 of the disk $|z| < \rho$ such that $f(z_0) \neq 0, \infty$. Then $f(z)$ is holomorphic and has no zero in a disk $|z - z_0| < \delta$ interior to the disk $|z| < \rho$. We are going to show that, in the disk $|z - z_0| < \delta$, we have

$$\log f(z) = \frac{1}{2\pi} \int_0^{2\pi} \log |f(\rho e^{i\varphi})| \frac{\rho e^{i\varphi} + z}{\rho e^{i\varphi} - z} d\varphi$$
$$- \sum_{\lambda=1}^h \log \frac{\rho^2 - \bar{a}_\lambda z}{\rho(z - a_\lambda)} + \sum_{\mu=1}^k \log \frac{\rho^2 - \bar{b}_\mu z}{\rho(z - b_\mu)} + ic , \tag{1.41}$$

where the logarithms are all holomorphic branches in the disk $|z - z_0| < \delta$ and c is a real constant. First note that the integral in (1.41) defines a function $I(z)$ holomorphic for $|z| < \rho$. In fact, from

$$\frac{I(z') - I(z)}{z' - z} - \frac{1}{2\pi} \int_0^{2\pi} \log |f(\rho e^{i\varphi})| \frac{2\rho e^{i\varphi}}{(\rho e^{i\varphi} - z)^2} d\varphi$$
$$= \frac{z' - z}{2\pi} \int_0^{2\pi} \log |f(\rho e^{i\varphi})| \frac{2\rho e^{i\varphi}}{(\rho e^{i\varphi} - z')(\rho e^{i\varphi} - z)^2} d\varphi ,$$

it follows that

$$I'(z) = \frac{1}{2\pi} \int_0^{2\pi} \log|f(\rho e^{i\varphi})| \frac{2\rho e^{i\varphi}}{(\rho e^{i\varphi} - z)^2} d\varphi .$$

Next note that both sides of (1.41) (the last term ic being excepted) are holomorphic functions in the disk $|z - z_0| < \delta$ and by (1.1) their real parts are equal, hence they differ only by a constant. Differentiating both sides of (1.41), we get,

$$\begin{aligned}
\frac{f'(z)}{f(z)} &= \frac{1}{2\pi} \int_0^{2\pi} \log|f(\rho e^{i\varphi})| \frac{2\rho e^{i\varphi}}{(\rho e^{i\varphi} - z)^2} d\varphi \\
&+ \sum_{\lambda=1}^{h} \frac{\rho^2 - |a_\lambda|^2}{(z - a_\lambda)(\rho^2 - \bar{a}_\lambda z)} - \sum_{\mu=1}^{k} \frac{\rho^2 - |b_\mu|^2}{(z - b_\mu)(\rho^2 - \bar{b}_\mu z)} .
\end{aligned} \tag{1.42}$$

This formula holds for $|z - z_0| < \delta$ and in particular for $z = z_0$. Since z_0 is arbitrary, (1.42) holds for $|z| < \rho$. (At a zero or pole of $f(z)$ in the disk $|z| < \rho$, both sides of (1.42) become ∞.)

Theorem 1.2. Let $f(z)$ be a meromorphic function in a domain $|z| < R$ $(0 < R \leq \infty)$ such that $c_0 = f(0) \neq 0, \infty$. Then for $0 < r < \rho < R$ we have

$$m\left(r, \frac{f'}{f}\right) < 4\log^+ T(\rho, f) + 3\log^+ \frac{1}{\rho - r} + 4\log^+ \rho + 2\log^+ \frac{1}{r}$$

$$+ 4\log^+ \log^+ \frac{1}{|c_0|} + 16 . \tag{1.43}$$

This theorem plays an important role in Nevanlinna's theory of meromorphic functions.

Proof. Consider a disk $|z| < \rho$ $(0 < \rho < R)$. By (1.42), in this disk, we have

$$\begin{aligned}
\left|\frac{f'(z)}{f(z)}\right| &\leq \frac{2\rho}{(\rho - |z|)^2} \frac{1}{2\pi} \int_0^{2\pi} |\log|f(\rho e^{i\varphi})|| d\varphi + \sum_{\lambda=1}^{h} \frac{\rho^2 - |a_\lambda|^2}{|z - a_\lambda||\rho^2 - \bar{a}_\lambda z|} \\
&+ \sum_{\mu=1}^{k} \frac{\rho^2 - |b_\mu|^2}{|z - b_\mu||\rho^2 - \bar{b}_\mu z|} .
\end{aligned}$$

Next from

$$|\rho^2 - \bar{a}_\lambda z| \geq \rho^2 - |a_\lambda||z| \geq \rho^2 - \rho|z| = \rho(\rho - |z|) \,,$$

$$\frac{\rho^2 - |a_\lambda|^2}{|z - a_\lambda||\rho^2 - \bar{a}_\lambda z|} = \frac{\rho(\rho^2 - |a_\lambda|^2)}{|\rho^2 - \bar{a}_\lambda z|^2}\left|\frac{\rho^2 - \bar{a}_\lambda z}{\rho(z - a_\lambda)}\right| \leq \frac{\rho}{(\rho - |z|)^2}\left|\frac{\rho^2 - \bar{a}_\lambda z}{\rho(z - a_\lambda)}\right| \,,$$

and

$$\frac{\rho^2 - |b_\mu|^2}{|z - b_\mu||\rho^2 - \bar{b}_\mu z|} \leq \frac{\rho}{(\rho - |z|)^2}\left|\frac{\rho^2 - \bar{b}_\mu z}{\rho(z - b_\mu)}\right| \,,$$

we have

$$\left|\frac{f'(z)}{f(z)}\right| \leq \frac{2\rho}{(\rho - |z|)^2}\left(\frac{1}{2\pi}\int_0^{2\pi} |\log|f(\rho e^{i\varphi})||d\varphi \right.$$
$$\left. + \sum_{\lambda=1}^h \left|\frac{\rho^2 - \bar{a}_\lambda z}{\rho(z - a_\lambda)}\right| + \sum_{\mu=1}^k \left|\frac{\rho^2 - \bar{b}_\mu z}{\rho(z - b_\mu)}\right|\right) \,. \tag{1.44}$$

By (1.16),

$$\frac{1}{2\pi}\int_0^{2\pi} |\log|f(\rho e^{i\varphi})||d\varphi = m(\rho, f) + m\left(\rho, \frac{1}{f}\right) \leq 2T(\rho, f) + \log\frac{1}{|c_0|} \,. \tag{1.45}$$

On the other hand, if we set

$$\chi(z, z_0) = \frac{\rho^2 - \bar{z}_0 z}{\rho(z - z_0)} \quad (|z_0| < \rho, z_0 \neq 0) \,, \tag{1.46}$$

then by (1.16),

$$m(r, \chi(z, z_0)) + N(r, \chi(z, z_0)) = m\left(r, \frac{1}{\chi(z, z_0)}\right) + N\left(r, \frac{1}{\chi(z, z_0)}\right)$$
$$+ \log\frac{\rho}{|z_0|} \quad (0 < r < \rho) \,.$$

Since

$$m\left(r, \frac{1}{\chi(z, z_0)}\right) = 0, \quad N\left(r, \frac{1}{\chi(z, z_0)}\right) = 0, \quad N(r, \chi(z, z_0)) = \log^+\frac{r}{|z_0|} \,,$$

we get

$$m(r, \chi(z, z_0)) = \log \frac{\rho}{|z_0|} - \log^+ \frac{r}{|z_0|} . \tag{1.47}$$

(1.44), (1.45) and (1.46) yield

$$\left| \frac{f'(z)}{f(z)} \right| \leq \frac{2\rho}{(\rho - |z|)^2} \left\{ 2T(\rho, f) + \log \frac{1}{|c_0|} + \sum_{\lambda=1}^{h} |\chi(z, a_\lambda)| + \sum_{\mu=1}^{k} |\chi(z, b_\mu)| \right\} .$$

In this inequality, setting $z = re^{i\theta}$, taking $\log^+$ of both sides and making use of (1.22) and (1.23), we find

$$\log^+ \left| \frac{f'(re^{i\theta})}{f(re^{i\theta})} \right| \leq \log^+ \frac{2\rho}{(\rho - r)^2} + \log^+ 2T(\rho, f) + \log^+ \log^+ \frac{1}{|c_0|}$$
$$+ \sum_{\lambda=1}^{h} \log^+ |\chi(re^{i\theta}, a_\lambda)| + \sum_{\mu=1}^{k} \log^+ |\chi(re^{i\theta}, b_\mu)|$$
$$+ \log \left\{ n\left(\rho, \frac{1}{f}\right) + n(\rho, f) + 2 \right\} .$$

Next integrating both sides with respect to θ from 0 to 2π, multiplying both sides by $1/2\pi$ and making use of (1.47), we find

$$m\left(r, \frac{f'}{f}\right) \leq \log^+ \frac{2\rho}{(\rho - r)^2} + \log^+ 2T(\rho, f) + \log^+ \log^+ \frac{1}{|c_0|}$$
$$+ N\left(\rho, \frac{1}{f}\right) - N\left(r, \frac{1}{f}\right) + N(\rho, f) - N(r, f)$$
$$+ \log \left\{ n\left(\rho, \frac{1}{f}\right) + n(\rho, f) + 2 \right\} . \tag{1.48}$$

For the sake of simplicity set

$$n(r) = n\left(r, \frac{1}{f}\right) + n(r, f) , \quad N(r) = N\left(r, \frac{1}{f}\right) + N(r, f) = \int_0^r \frac{n(t)}{t} dt .$$
$$(0 < r < R)$$

Taking a value ρ_1 such that $0 < r < \rho_1 < \rho < R$ and setting $\rho = \rho_1$ in (1.48), we obtain

$$m\left(r, \frac{f'}{f}\right) \leq \log^+ \frac{2\rho_1}{(\rho_1 - r)^2} + \log^+ 2T(\rho_1, f) + \log^+ \log^+ \frac{1}{|c_0|}$$
$$+ N(\rho_1) - N(r) + \log(n(\rho_1) + 2) . \tag{1.49}$$

In what follows, we are going to find upper bounds of $n(\rho_1)$ and $N(\rho_1) - N(r)$. First from

$$n(\rho_1)\log\frac{\rho}{\rho_1} \leq \int_{\rho_1}^{\rho}\frac{n(t)}{t}dt \leq N(\rho) \leq T(\rho,f) + T\left(\rho,\frac{1}{f}\right)$$

$$= 2T(\rho,f) + \log\frac{1}{|c_0|} \, ,$$

$$\log\frac{\rho}{\rho_1} = -\log\left(1 - \frac{\rho - \rho_1}{\rho}\right) \geq \frac{\rho - \rho_1}{\rho}$$

we have

$$n(\rho_1) \leq \frac{\rho}{\rho - \rho_1}\left\{2T(\rho,f) + \log^+\frac{1}{|c_0|}\right\} \, ,$$

$$\log^+ n(\rho_1) \leq \log\frac{\rho}{\rho - \rho_1} + \log^+ 2T(\rho,f) + \log^+\log^+\frac{1}{|c_0|} + \log 2 \, . \tag{1.50}$$

Next, $N(r)$ being a convex function of $\log r$, we have

$$N(\rho_1) - N(r) \leq \frac{\log\frac{\rho_1}{r}}{\log\frac{\rho}{r}}\{N(\rho) - N(r)\} \leq \frac{\log\frac{\rho_1}{r}}{\log\frac{\rho}{r}}N(\rho)$$

$$\leq \frac{\log\frac{\rho_1}{r}}{\log\frac{\rho}{r}}\left\{2T(\rho,f) + \log^+\frac{1}{|c_0|}\right\}$$

which together with the inequalities

$$\log\frac{\rho_1}{r} = \log\left(1 + \frac{\rho_1 - r}{r}\right) \leq \frac{\rho_1 - r}{r}, \quad \log\frac{\rho}{r} \geq \frac{\rho - r}{\rho}$$

yield

$$N(\rho_1) - N(r) \leq \frac{\rho}{r}\frac{\rho_1 - r}{\rho - r}\left\{2T(\rho,f) + \log^+\frac{1}{|c_0|}\right\} \, . \tag{1.51}$$

In this inequality we only assume $r < \rho_1 < \rho$. Taking in particular

$$\rho_1 - r = \frac{r(\rho - r)}{2\left\{T(\rho,f) + \log^+\frac{1}{|c_0|} + 1\right\}\rho + 1} \, , \tag{1.52}$$

(1.51) becomes

$$N(\rho_1) - N(r) < 1 \, . \tag{1.53}$$

It remains to find upper bounds of $1/(\rho_1 - r)$ and $1/(\rho - \rho_1)$. First we have

$$\frac{1}{\rho_1 - r} = \frac{1}{\rho - r}\frac{1}{r}\left\{2[T(\rho, f) + \log^+ \frac{1}{|c_0|} + 1]\rho + 1\right\} ,$$

$$\log^+ \frac{1}{\rho_1 - r} \leq \log^+ \frac{1}{\rho - r} + \log^+ \frac{1}{r} + \log^+ \rho + \log^+ T(\rho, f)$$
$$+ \log^+ \log^+ \frac{1}{|c_0|} + 2\log 2 + \log 3 . \tag{1.54}$$

Next we have

$$\rho - \rho_1 = (\rho - r) - (\rho_1 - r) = (1 - \lambda)(\rho - r) ,$$

where

$$\lambda = \frac{r}{2\{T(\rho, f) + \log^+ \frac{1}{|c_0|} + 1\}\rho + 1} < \frac{1}{2} ,$$

and hence

$$\rho - \rho_1 > \frac{1}{2}(\rho - r) ,$$
$$\log^+ \frac{1}{\rho - \rho_1} \leq \log^+ \frac{1}{\rho - r} + \log 2 . \tag{1.55}$$

Finally from (1.49), (1.50), (1.53), (1.54) and (1.55), we get

$$m\left(r, \frac{f'}{f}\right) < \log^+ \rho_1 + 2\log^+ \frac{1}{\rho_1 - r} + \log^+ T(\rho_1, f) + \log^+ \log^+ \frac{1}{|c_0|}$$
$$+ 1 + \log^+ n(\rho_1) + 4\log 2$$
$$\leq \log^+ \rho_1 + 2\log^+ \frac{1}{\rho_1 - r} + \log \frac{\rho}{\rho - \rho_1} + 2\log^+ T(\rho, f)$$
$$+ 2\log^+ \log^+ \frac{1}{|c_0|} + 1 + 6\log 2$$
$$\leq 4\log^+ \rho + 3\log^+ \frac{1}{\rho - r} + 2\log^+ \frac{1}{r} + 4\log^+ T(\rho, f)$$
$$+ 4\log^+ \log^+ \frac{1}{|c_0|} + 11\log 2 + 2\log 3 + 1 .$$

Hence we have (1.43).

In Theorem 1.2, it is assumed $f(0) \neq 0, \infty$. If the point $z = 0$ is a zero or pole of the function $f(z)$, supposed non-identically equal to zero, then in the neighborhood of the point $z = 0$,

$$f(z) = c_s z^s + c_{s+1} z^{s+1} + \dots \quad (c_s \neq 0) .$$

We can apply Theorem 1.2 to the function $f_1(z) = z^{-s} f(z)$ and get, for $0 < r < \rho < R$, the inequality

$$m\left(r, \frac{f_1'}{f_1}\right) < 4\log^+ T(\rho, f_1) + 3\log^+ \frac{1}{\rho - r} + 4\log^+ \rho + 2\log^+ \frac{1}{r}$$
$$+ 4\log^+ \log^+ \frac{1}{|c_s|} + 16 . \tag{1.56}$$

To get an upper bound of $m(r, \frac{f'}{f})$, first from

$$\frac{f_1'(z)}{f_1(z)} = -\frac{s}{z} + \frac{f'(z)}{f(z)} ,$$

we have

$$m\left(r, \frac{f'}{f}\right) \leq m\left(r, \frac{f_1'}{f_1}\right) + \log^+ |s| + \log^+ \frac{1}{r} + \log 2 . \tag{1.57}$$

Next from

$$m(r, f_1) \leq m(r, z^{-s}) + m(r, f) ,$$

$$m(r, z^{-s}) = \begin{cases} s\log^+ \frac{1}{r} & \text{if } s > 0 \\ -s\log^+ r & \text{if } s < 0 , \end{cases}$$

$$N(r, f_1) = N(r, f) - n(0, f)\log r ,$$

where $n(0, f)$ is equal to 0 or $-s$, according to $s > 0$ or $s < 0$, we have

$$T(r, f_1) \leq T(r, f) + |s|\left(\log^+ r + \log^+ \frac{1}{r}\right) . \tag{1.58}$$

(1.56), (1.57) and (1.58) yield

$$m\left(r, \frac{f'}{f}\right) < 4\log^+ T^+(\rho, f) + 3\log^+ \frac{1}{\rho - r} + 8\log^+ \rho + 6\log^+ \frac{1}{r}$$
$$+ 4\log^+ \log^+ \frac{1}{|c_s|} + 5\log^+ |s| + 9\log 2 + 16 ,$$

where $T^{+}(\rho, f) = \max\{T(\rho, f), 0\}$. Evidently this inequality also holds, when $f(0) \neq 0, \infty$. We have therefore the following corollary:

Corollary 1.1. Let $f(z)$ be a meromorphic function non-identically equal to zero in a domain $|z| < R$ $(0 < R \leq \infty)$. Then for $0 < r < \rho < R$ we have

$$m\left(r, \frac{f'}{f}\right) < 4\log^{+} T^{+}(\rho, f) + 3\log^{+}\frac{1}{\rho - r} + 8\log^{+}\rho + 6\log^{+}\frac{1}{r}$$

$$+ 4\log^{+}\log^{+}\frac{1}{|c_s|} + 5\log^{+}|s| + 25 . \tag{1.59}$$

Sometimes we need the following generalization of Corollary 1.1:

Theorem 1.3. Let $f(z)$ be a meromorphic function non-identically equal to zero and $n \geq 1$ a positive integer. Then there are positive constants A, B, C, D such that for $1 \leq r < \rho$ we have

$$m\left(r, \frac{f^{(n)}}{f}\right) < A\log^{+} T(\rho, f) + B\log\rho + C\log^{+}\frac{1}{\rho - r} + D . \tag{1.60}$$

Proof. By Corollary 1.1, Theorem 1.3 holds when $n = 1$. Now suppose that Theorem 1.3 is true for a positive integer n. We are going to show that it is also true for $n + 1$. For this purpose, distinguish two cases:

$1° f^{(n)}(z) \equiv 0$. In this case, obviously there are positive constants A_1, B_1, C_1, D_1 such that for $1 \leq r < \rho$ we have

$$m\left(r, \frac{f^{(n+1)}}{f}\right) < A_1\log^{+} T(\rho, f) + B_1\log\rho + C_1\log^{+}\frac{1}{\rho - r} + D_1 . \tag{1.61}$$

$2° f^{(n)}(z) \not\equiv 0$. In this case, we first deduce from the identity

$$\frac{f^{(n+1)}(z)}{f(z)} = \frac{f^{(n+1)}(z)}{f^{(n)}(z)}\frac{f^{(n)}(z)}{f(z)} ,$$

the inequality

$$m\left(r, \frac{f^{(n+1)}}{f}\right) \leq m\left(r, \frac{f^{(n+1)}}{f^{(n)}}\right) + m\left(r, \frac{f^{(n)}}{f}\right) . \tag{1.62}$$

Next, by hypothesis, (1.60) holds and, on the other hand, by Corollary 1.1, we have, for $1 \leq r < \rho$,

$$m\left(r, \frac{f^{(n+1)}}{f^{(n)}}\right) < 4\log^+ T(\rho, f^{(n)}) + 8\log^+ \rho + 3\log^+ \frac{1}{\rho - r} + d , \quad (1.63)$$

where d is a positive constant. Now let $1 \leq r < \rho$ and set $\rho_1 = (r + \rho)/2$, then

$$m\left(r, \frac{f^{(n+1)}}{f^{(n)}}\right) < 4\log^+ T(\rho_1, f^{(n)}) + 8\log^+ \rho_1 + 3\log^+ \frac{1}{\rho_1 - r} + d . \quad (1.64)$$

Since

$$\begin{aligned}
T(\rho_1, f^{(n)}) = m(\rho_1, f^{(n)}) + N(\rho_1, f^{(n)}) &\leq m(\rho_1, f) + m\left(\rho_1, \frac{f^{(n)}}{f}\right) \\
&\quad + (n+1)N(\rho_1, f) \\
&\leq (n+1)T(\rho_1, f) + m\left(\rho_1, \frac{f^{(n)}}{f}\right)
\end{aligned}$$
$$(1.65)$$

and by (1.60),

$$m\left(\rho_1, \frac{f^{(n)}}{f}\right) < A\log^+ T(\rho, f) + B\log \rho + C\log^+ \frac{1}{\rho - \rho_1} + D , \quad (1.66)$$

we see finally from (1.62), (1.60), (1.64), (1.65) and (1.66) that there are positive constants A_1, B_1, C_1, D_1 such that (1.61) holds for $1 \leq r < \rho$.

In order to get some estimates of $m(r, f^{(n)}/f)$, which are convenient for certain applications, we need the following lemma due to Borel:

Lemma 1.1. Let $\varphi(x)$ be a continuous non-decreasing function for $x > 0$, tending to ∞ with x. Then the inequality

$$\varphi\left\{x + \frac{1}{\log \varphi(x)}\right\} < \{\varphi(x)\}^2 \quad (1.67)$$

holds, when x is exterior to a sequence of intervals of finite total length.

Proof. If there is a positive number a such that (1.67) holds for $x \geq a$, then the conclusion of Lemma 1.1 is evident. We may therefore assume that the inequality

$$\varphi\left\{x + \frac{1}{\log \varphi(x)}\right\} \geq \{\varphi(x)\}^2 \quad (1.68)$$

is satisfied by arbitrarily large values of x. Let x_0 be a value satisfying $\varphi(x_0) > 1$ and (1.68). Consider the sequence of values

$$x_0, x_1 = x_0 + \frac{1}{\log \varphi(x_0)}, \quad x_2 = x_1 + \frac{1}{\log \varphi(x_1)}, \dots,$$

$$x_n = x_{n-1} + \frac{1}{\log \varphi(x_{n-1})}, \dots. \tag{1.69}$$

Obviously $x_n (n = 0, 1, 2, \dots)$ is an increasing sequence. We are going to show that not all terms of this sequence satisfy (1.68). In fact, assume, on the contrary, that

$$\varphi \left\{ x_n + \frac{1}{\log \varphi(x_n)} \right\} \geq \{\varphi(x_n)\}^2 \quad (n = 0, 1, 2, \dots). \tag{1.70}$$

Then

$$\log \varphi(x_{n+1}) \geq 2 \log \varphi(x_n) \quad (n = 0, 1, 2, \dots) \tag{1.71}$$

and we have successively

$$\log \varphi(x_1) \geq 2 \log \varphi(x_0),$$
$$\log \varphi(x_2) \geq 2^2 \log \varphi(x_0),$$
$$\dots$$
$$\log \varphi(x_n) \geq 2^n \log \varphi(x_0),$$

which shows that $\varphi(x_n)$ tends to ∞ with n. On the the other hand,

$$x_n = x_0 + \sum_{j=0}^{n-1} \frac{1}{\log \varphi(x_j)} \leq x_0 + \frac{1}{\log \varphi(x_0)} \sum_{j=0}^{n-1} \frac{1}{2^j},$$

which shows that x_n is bounded. We get a contradiction.

Let x_{m_0} be the first term of the sequence $x_n (n = 0, 1, 2, \dots)$ satisfying (1.67), obviously $m_0 \geq 1$. Set $X = x_{m_0}$, then

$$\log \varphi(X) \geq 2^{m_0} \log \varphi(x_0), \tag{1.72}$$

$$X - x_0 \leq \frac{1}{\log \varphi(x_0)} \sum_{j=0}^{m_0-1} \frac{1}{2^j}. \tag{1.73}$$

Since X satisfies (1.67), by continuity there is an interval (X, x_0') in which (1.67) holds and such that x_0' satisfies (1.68). By (1.72),

$$\log \varphi(x_0') \geq 2^{m_0} \log \varphi(x_0) . \tag{1.74}$$

Starting from x_0' and repeating the same process, we get again a value $X' = x_{m_1}' > x_0'(m_1 \geq 1)$ satisfying (1.67) and such that

$$\log \varphi(X') \geq 2^{m_1} \log \varphi(x_0') , \tag{1.75}$$

$$X' - x_0' \leq \frac{1}{\log \varphi(x_0')} \sum_{j=0}^{m_1-1} \frac{1}{2^j} . \tag{1.76}$$

Since X' satisfies (1.67), there is an interval (X', x_0'') in which (1.67) holds and such that x_0'' satisfies (1.68). By (1.75),

$$\log \varphi(x_0'') \geq 2^{m_1} \log \varphi(x_0') . \tag{1.77}$$

In this way we get successively a sequence of intervals

$$(x_0, X), (x_0', X'), (x_0'', X''), \ldots, (x_0^{(p)}, X^{(p)}), \ldots \tag{1.78}$$

such that (1.67) holds in the intervals

$$[X, x_0'), [X', x_0''), \ldots, [X^{(p)}, x_0^{(p+1)}), \ldots$$

and that we have in general

$$X^{(p)} - x_0^{(p)} \leq \frac{1}{\log \varphi(x_0^{(p)})} \sum_{j=0}^{m_p-1} \frac{1}{2^j} , \tag{1.79}$$

$$\log \varphi(x_0^{(p+1)}) \geq 2^{m_p} \log \varphi(x_0^{(p)}) , \tag{1.80}$$

where $m_p \geq 1$. We then have successively

$$\log \varphi(x_0') \geq 2^{m_0} \log \varphi(x_0) ,$$
$$\log \varphi(x_0'') \geq 2^{m_0+m_1} \log \varphi(x_0) ,$$
$$\ldots$$
$$\log \varphi(x_0^{(p)}) \geq 2^{m_0+m_1+\ldots+m_{p-1}} \log \varphi(x_0) ,$$

which implies that $x_0^{(p)}$ tends to ∞ with p. Next we have

$$X - x_0 \le \frac{1}{\log \varphi(x_0)} \sum_{j=0}^{m_0-1} \frac{1}{2^j} \ ,$$

$$X' - x_0' \le \frac{1}{\log \varphi(x_0)} \sum_{j=m_0}^{m_0+m_1-1} \frac{1}{2^j} \ ,$$

$$\cdots$$

$$X^{(p)} - x_0^{(p)} \le \frac{1}{\log \varphi(x_0)} \sum_{j=m_0+m_1+\ldots+m_{p-1}}^{m_0+m_1+\ldots+m_p-1} \frac{1}{2^j} \ .$$

Consequently the total length of the sequence of intervals (1.78) does not exceed

$$\frac{1}{\log \varphi(x_0)} \sum_{j=0}^{\infty} \frac{1}{2^j} = \frac{2}{\log \varphi(x_0)} \ .$$

Corollary 1.2. Let $f(z)$ be a non-constant meromorphic function and $n \ge 1$ a positive integer. Then the following estimates of $m(r, f^{(n)}/f)$ hold:

1° If the order of $f(z)$ is finite, then

$$m\left(r, \frac{f^{(n)}}{f}\right) = O(\log r) \ . \tag{1.81}$$

2° In the general case, there is a sequence of intervals $\{I_p\}$ of finite total length and depending only on $f(z)$, such that when r is exterior to $\{I_p\}$, we have

$$m\left(r, \frac{f^{(n)}}{f}\right) = O\{\log T(r, f) + \log r\} \ . \tag{1.82}$$

Proof. If the order λ of $f(z)$ is finite, then

$$T(r, f) < r^{\lambda+1} \ , \tag{1.83}$$

when r is sufficiently large. In (1.60) taking $\rho = 2r$ and making use of (1.83), we see that when r is sufficiently large,

$$m\left(r, \frac{f^{(n)}}{r}\right) < K \log r \ ,$$

where K is a positive constant.

In the general case, $T(r, f)$ is continuous, non-decreasing for $r > 0$ and tends to ∞ with r. We can then apply Lemma 1.1 to the function $\varphi(r) = T(r, f)$. Consequently there is a sequence of intervals $\{I_p\}$ of finite total length such that we have

$$\varphi \left\{ r + \frac{1}{\log \varphi(r)} \right\} < \{\varphi(r)\}^2 , \qquad (1.84)$$

when r is exterior to $\{I_p\}$. Let r be exterior to $\{I_p\}$ and in (1.60) take $\rho = r + 1/\log \varphi(r)$. Then making use of (1.84), we see that

$$m \left(r, \frac{f^{(n)}}{f} \right) < K'\{\log T(r, f) + \log r\} ,$$

when r is sufficiently large and exterior to $\{I_p\}$, where K' is a positive constant.

Remark. Consider a non-constant meromorphic function $f(z)$, a finite value a and a positive integer n. By Theorem 1.3, for $1 \le r < \rho$, we have

$$m \left(r, \frac{f^{(n)}}{f - a} \right) < A' \log^+ T(\rho, f - a) + B' \log \rho + C' \log^+ \frac{1}{\rho - r} + D' ,$$

where A', B', C', and D' are positive constants. Since

$$T(\rho, f - a) \le T(\rho, f) + \log^+ |a| + \log 2 ,$$

we have also

$$m \left(r, \frac{f^{(n)}}{f - a} \right) < A' \log^+ T(\rho, f) + B' \log \rho + C' \log^+ \frac{1}{\rho - r} + D'' . \quad (1.85)$$

On the basis of (1.85), evidently the two estimates (1.81) and (1.82) of $m(r, f^{(n)}/f)$ in Corollary 1.2 are also true for $m(r, f^{(n)}/(f - a))$.

1.6. SECOND FUNDAMENTAL THEOREM

Theorem 1.4. Let $f(z)$ be a non-constant meromorphic function and $a_j (j = 1, 2, \ldots, q; q \ge 2)$ be q distinct finite values. Then for $r > 0$, we have

$$\sum_{j=1}^{q} m \left(r, \frac{1}{f - a_j} \right) + m(r, f) \le 2T(r, f) - N_1(r) + S(r) , \qquad (1.86)$$

where

$$N_1(r) = \{2N(r, f) - N(r, f')\} + N\left(r, \frac{1}{f'}\right) \qquad (1.87)$$

and $S(r)$ satisfies the following conditions:

1° If the order of $f(z)$ is finite, then

$$S(r) = O(\log r) . \qquad (1.88)$$

2° In the general case, there is a sequence of intervals $\{I_p\}$ of finite total length and depending only on $f(z)$, such that when r is exterior to $\{I_p\}$, we have

$$S(r) = O\{\log T(r, f) + \log r\} . \qquad (1.89)$$

This theorem is one form of the second fundamental theorem.

Proof. The method of proof is to introduce the auxiliary function

$$F(z) = \sum_{j=1}^{q} \frac{1}{f(z) - a_j}$$

and to find a lower bound and an upper bound of $m(r, F)$. To find a lower bound of $m(r, F)$, consider a value $r > 0$. Set

$$\delta = \min\{1, |a_j - a_k| \ (1 \le j, k \le q, j \ne k)\}$$

and define E_j to be the set of values φ of the interval $0 \le \varphi \le 2\pi$, satisfying the inequality

$$|f(re^{i\varphi}) - a_j| < \frac{\delta}{2q} .$$

If $z = re^{i\varphi}$ with $\varphi \in E_j$, then

$$F(z) = \frac{1}{f(z) - a_j} \left\{ 1 + \sum_{k \ne j} \frac{f(z) - a_j}{f(z) - a_k} \right\} ,$$

$$|f(z) - a_k| \ge |a_j - a_k| - |f(z) - a_j| > \delta - \frac{\delta}{2q} \ge \frac{3}{4}\delta \quad (k \ne j) ,$$

$$\sum_{k \ne j} \left| \frac{f(z) - a_j}{f(z) - a_k} \right| < q\frac{2}{3q} = \frac{2}{3} ,$$

$$|F(z)| > \frac{1}{3}\left| \frac{1}{f(z) - a_j} \right| ,$$

hence

$$\log^+ |F(re^{i\varphi})| \geq \log^+ \left| \frac{1}{f(re^{i\varphi}) - a_j} \right| - \log 3 \quad (\varphi \in E_j) .$$

Noting that the sets E_j $(j = 1, 2, \ldots, q)$ are mutually disjoint, we have

$$m(r, F) \geq \frac{1}{2\pi} \sum_{j=1}^{q} \int_{E_j} \log^+ |F(re^{i\varphi})| d\varphi$$

$$\geq \frac{1}{2\pi} \sum_{j=1}^{q} \int_{E_j} \log^+ \left| \frac{1}{f(re^{i\varphi}) - a_j} \right| d\varphi - \log 3 .$$

Denoting by H_j the complement of E_j with respect to the interval $0 \leq \varphi \leq 2\pi$, then

$$\frac{1}{2\pi} \int_{E_j} \log^+ \left| \frac{1}{f(re^{i\varphi}) - a_j} \right| d\varphi = m \left(r, \frac{1}{f - a_j} \right)$$

$$- \frac{1}{2\pi} \int_{H_j} \log^+ \left| \frac{1}{f(re^{i\varphi}) - a_j} \right| d\varphi$$

$$\geq m \left(r, \frac{1}{f - a_j} \right) - \log \frac{2q}{\delta} ,$$

hence

$$m(r, F) \geq \sum_{j=1}^{q} m \left(r, \frac{1}{f - a_j} \right) - q \log \frac{2q}{\delta} - \log 3 . \qquad (1.90)$$

Now to find an upper bound of $m(r, F)$, we write

$$F(z) = \frac{1}{f'(z)} \sum_{j=1}^{q} \frac{f'(z)}{f(z) - a_j}$$

and obtain

$$m(r, F) \leq m \left(r, \frac{1}{f'} \right) + \sum_{j=1}^{q} m \left(r, \frac{f'}{f - a_j} \right) + \log q .$$

Since

$$m \left(r, \frac{1}{f'} \right) = m(r, f') + N(r, f') - N \left(r, \frac{1}{f'} \right) + \log \frac{1}{|c|}$$

by (1.16), where $c \neq 0$ is a constant, and

$$m(r, f') \leq m(r, f) + m\left(r, \frac{f'}{f}\right) ,$$

it follows that

$$m(r, F) \leq m(r, f) + N(r, f') - N\left(r, \frac{1}{f'}\right)$$

$$+ m\left(r, \frac{f'}{f}\right) + \sum_{j=1}^{q} m\left(r, \frac{f'}{f - a_j}\right) + \log \frac{1}{|c|} + \log q . \tag{1.91}$$

Inequalities (1.90) and (1.91) yield

$$\sum_{j=1}^{q} m\left(r, \frac{1}{f - a_j}\right) + m(r, f) \leq 2T(r, f) - N_1(r) + S(r) , \tag{1.92}$$

where

$$S(r) = m\left(r, \frac{f'}{f}\right) + \sum_{j=1}^{q} m\left(r, \frac{f'}{f - a_j}\right) + \alpha ,$$

α being a constant. By Corollary 1.2 and Remark, evidently $S(r)$ satisfies the conditions 1° and 2° in Theorem 1.4.

Now we are going to state the second fundamental theorem in another form. For this, it is convenient to use the notations:

$$N(r, \infty) = N(r, f), \quad N(r, a) = N\left(r, \frac{1}{f - a}\right) \quad (a \text{ finite}) \tag{1.93}$$

introduced by Nevanlinna.

Theorem 1.5. Let $f(z)$ be a non-constant meromorphic function and $a_j (j = 1, 2, \ldots, q; q \geq 3) q$ distinct values, finite or infinite. Then for $r > 0$, we have

$$(q - 2)T(r, f) \leq \sum_{j=1}^{q} N(r, a_j) - N_1(r) + S(r) , \tag{1.94}$$

where $N_1(r)$ is defined by (1.87) and $S(r)$ satisfies the conditions 1° and 2° in Theorem 1.4.

Proof. We consider only the case that one of the values $a_j (j = 1, 2, \ldots, q)$ is ∞, for instance $a_q = \infty$. Then by Theorem 1.4, we have

$$\sum_{j=1}^{q-1} m\left(r, \frac{1}{f - a_j}\right) + m(r, f) \leq 2T(r, f) - N_1(r) + S(r) .$$

To both sides of this inequality adding the sum $\Sigma_{j=1}^{q} N(r, a_j)$, and then noting that

$$T(r, f) \leq T\left(r, \frac{1}{f - a_j}\right) + k_j \quad (j = 1, 2, \ldots, q - 1) \,,$$

by (1.36), where $k_j (j = 1, 2, \ldots, q - 1)$ are positive constants, we get

$$(q - 2)T(r, f) \leq \sum_{j=1}^{q} N(r, a_j) - N_1(r) + S_1(r) \,,$$

where $S_1(r) = S(r) + \sum_{j=1}^{q-1} k_j$ evidently also satisfies the conditions $1°$ and $2°$ in Theorem 1.4. The case that the values $a_j (j = 1, 2, \ldots, q)$ are all finite, is treated by the same method.

Now let us study the term $N_1(r)$ defined by (1.87). $N_1(r)$ consists of two parts $2N(r, f) - N(r, f')$ and $N(r, 1/f')$. They are both non-negative for $r \geq 1$. Consider first the second part $N(r, 1/f')$. This part is related to the points z_0 such that $f(z_0)$ is finite and z_0 is a zero of order greater than one of the function $f(z) - f(z_0)$. For simplicity, let us name such a point z_0 a multiple point of the first kind of $f(z)$ and the order of z_0 as a zero of $f(z) - f(z_0)$ the order of z_0. Evidently $n(t, 1/f')$ is equal to the number of multiple points of the first kind of $f(z)$ in the disk $|z| \leq t$, each one of such points being counted as many times as its order minus one. Consider now the first part $2N(r, f) - N(r, f')$. This part is related to poles of order greater than one of $f(z)$, namely multiple poles of $f(z)$. Evidently $2n(t, f) - n(t, f')$ is equal to the number of multiple poles of $f(z)$ in the disk $|z| \leq t$, each multiple pole being counted as many times as its order minus one. Thus if we denote by $n_1(t)$ the number of multiple points (those of the first kind and multiple poles) of $f(z)$ in the disk $|z| \leq t$, each multiple point being counted as many times as its order minus one, then

$$n_1(t) = \{2n(t, f) - n(t, f')\} + n\left(t, \frac{1}{f'}\right) \,. \tag{1.95}$$

Consequently we have the formula

$$N_1(r) = \int_0^r \frac{n_1(t) - n_1(0)}{t} dt + n_1(0) \log r \,. \tag{1.96}$$

The notation $N(r, a)$ (a finite or infinite) introduced above may be expressed as

$$N(r, a) = \int_0^r \frac{n(t, a) - n(0, a)}{t} dt + n(0, a) \log r , \qquad (1.97)$$

where the meaning of $n(t, a)$ is self-evident. Nevanlinna also introduced the following notation:

$$\overline{N}(r, a) = \int_0^r \frac{\overline{n}(t, a) - \overline{n}(0, a)}{t} dt + \overline{n}(0, a) \log r , \qquad (1.98)$$

where $\overline{n}(t, a)$ denotes the number of the roots in the disk $|z| \leq t$ of the equation $f(z) = a$, each root being counted once. Noting that for any q distinct values, $a_j (j = 1, 2, \ldots , q)$ finite or infinite, we have

$$\sum_{j=1}^q n(t, a_j) - n_1(t) \leq \sum_{j=1}^q \overline{n}(t, a_j) ,$$

we deduce from Theorem 1.5 and (1.95) the following theorem:

Theorem 1.6. Let $f(z)$ be a non-constant meormorphic function and $a_j (j = 1, 2, \ldots , q; q \geq 3) q$ distinct values finite or infinite. Then for $r > 0$ we have

$$(q - 2) T(r, f) \leq \sum_{j=1}^q \overline{N}(r, a_j) + S(r) , \qquad (1.99)$$

where $S(r)$ satisfies the conditions $1°$ and $2°$ in Theorem 1.4.

In what follows, we give some applications of the Theorems 1.4, 1.5 and 1.6.

Corollary 1.3. Let $f(z)$ be a transcendental meromorphic function. Then for each value a finite or infinite, the equation $f(z) = a$ has an infinite number of roots, except for at most two exceptional values.

This Corollary is Picard theorem for meromorphic functions. An exceptional value a, if it exists, is called a Picard exceptional value of $f(z)$.

Proof. In the particular case $q = 3$, Theorem 1.5 yields the inequality

$$T(r, f) \leq \sum_{j=1}^3 N(r, a_j) + S(r) . \qquad (1.100)$$

Now suppose that for three values $a_j (j = 1, 2, 3)$, the equations $f(z) = a_j (j = 1, 2, 3)$ all have at most a finite number of roots. Then evidently

$$N(r, a_j) = O(\log r) \quad (j = 1, 2, 3) .$$

Consequently by the condition $2°$ in Theorem 1.4, when r is exterior to a sequence of intervals $\{I_p\}$ of finite total length and is sufficiently large, we have

$$T(r, f) \leq K\{\log T(r, f) + \log r\} ,$$

where K is a positive constant. But this is impossible, by (1.38).

Corollary 1.4. Let $f(z)$ be a transcendental meromorphic function of finite positive order ρ $(0 < \rho < \infty)$. Then for each value a finite or infinite, we have

$$\varlimsup_{r \to \infty} \frac{\log n(r, a)}{\log r} = \rho , \tag{1.101}$$

except for at most two exceptional values.

This Corollary is Borel's theorem for meromorphic functions. An exceptional value a, if it exists, is called a Borel exceptional value of $f(z)$.

Proof. Noting first that, for $r \geq 1$, we have

$$n(r, a) \log 2 \leq \int_r^{2r} \frac{n(t, a)}{t} dt \leq N(2r, a) ,$$

and then by the first fundamental theorem,

$$n(r, a) \log 2 \leq T(2r, f) + k ,$$

where k is a positive constant. So we have

$$\varlimsup_{r \to \infty} \frac{\log n(r, a)}{\log r} \leq \rho . \tag{1.102}$$

Now suppose that there are three values $a_j (j = 1, 2, 3)$ which do not satisfy (1.101). Then we can find a constant $\lambda(0 < \lambda < \rho)$ such that

$$n(r, a_j) < r^\lambda \quad (r \geq r_0) \quad (j = 1, 2, 3) .$$

Hence

$$N(r, a_j) - N(r_0, a_j) = \int_{r_0}^r \frac{n(t, a_j)}{t} dt < \int_{r_0}^r \frac{t^\lambda}{t} dt = \frac{1}{\lambda}(r^\lambda - r_0^\lambda) . \tag{1.103}$$

From (1.100), (1.103) and the relation $S(r) = O(\log r)$, it follows that, for sufficiently large values of r,

$$T(r,f) < hr^\lambda \, ,$$

where h is a positive constant. But this is impossible because $\lambda < \rho$.

Consider a transcendental meromorphic function $f(z)$ and a finite value a. By (1.36) we may write

$$m\left(r, \frac{1}{f-a}\right) = T(r,f) - N\left(r, \frac{1}{f-a}\right) + O(1) \, .$$

Dividing both sides of this equality by $T(r,f)$ and then taking lower limit, we get

$$\varliminf_{r\to\infty} \frac{m\left(r, \frac{1}{f-a}\right)}{T(r,f)} = 1 - \varlimsup_{r\to\infty} \frac{N\left(r, \frac{1}{f-a}\right)}{T(r,f)} \, . \tag{1.104}$$

We have also

$$\varliminf_{r\to\infty} \frac{m(r,f)}{T(r,f)} = 1 - \varlimsup_{r\to\infty} \frac{N(r,f)}{T(r,f)} \, . \tag{1.105}$$

Nevanlinna introduced the notation $\delta(a,f)$ defined for a finite or infinite as follows:

$$\delta(a,f) = 1 - \varlimsup_{r\to\infty} \frac{N(r,a)}{T(r,f)} \, . \tag{1.106}$$

Evidently

$$0 \le \delta(a,f) \le 1 \, .$$

Corollary 1.5. Let $f(z)$ be a transcendental meromorphic function and $a_j (j = 1, 2, \ldots, q; q \ge 2) q$ distinct finite values. Then

$$\sum_{j=1}^{q} \delta(a_j, f) + \delta(\infty, f) \le 2 \, . \tag{1.107}$$

Proof. By (1.86), we have

$$\sum_{j=1}^{q} \frac{1}{T(r,f)} m\left(r, \frac{1}{f-a_j}\right) + \frac{m(r,f)}{T(r,f)} \le 2 + \frac{S(r)}{T(r,f)} \, .$$

Taking lower limit and making use of (1.104) and (1.105), we get

$$\sum_{j=1}^{q} \delta(a_j, f) + \delta(\infty, f) \le 2 + \varlimsup_{r \to \infty} \frac{S(r)}{T(r, f)} \ .$$

Since $S(r)$ satisfies the condition $2°$ in Theorem 1.4, we have

$$\varlimsup_{r \to \infty} \frac{S(r)}{T(r, f)} \le \varlimsup_{\substack{r \to \infty \\ r \notin \{I_p\}}} \frac{S(r)}{T(r, f)} = 0$$

and (1.107).

If a value a is such that $\delta(a, f) > 0$, then a is called a deficient value of $f(z)$ or a Nevanlinna exceptional value of $f(z)$, and $\delta(a, f)$ the deficiency corresponding to the value a. It is easy to see that $f(z)$ can have at most countably many deficient values. In fact if we denote by σ_k the set of the deficient values satisfying the inequality $1/(k + 1) < \delta(a, f) \le 1/k$, and by σ the set of all the deficient values, then

$$\sigma = \bigcup_{k=1}^{\infty} \sigma_k$$

and, by (1.107), σ_k consists of at most $2(k + 1)$ deficient values. We have

$$\sum_{a} \delta(a, f) \le 2 \ , \tag{1.108}$$

where the summation is taken with respect to all the deficient values of $f(z)$. In particular for a transcendental entire function $f(z)$, we have $\delta(\infty, f) = 1$ and hence

$$\sum_{a \ne \infty} \delta(a, f) \le 1 \ . \tag{1.109}$$

Now we introduce another important notion that of completely multiple value. Consider a transcendental meromorphic function $f(z)$. A finite value a is called a completely multiple value of $f(z)$, if each zero of $f(z) - a$ has an order greater than one. ∞ is called a completely multiple value of $f(z)$, if each pole of $f(z)$ has an order greater than one.

Corollary 1.6. Let $f(z)$ be a transcendental meromorphic function. Then $f(z)$ has at most four completely multiple values.

Proof. Assume that $f(z)$ has five completely multiple values $a_j (j = 1, 2, 3, 4, 5)$. In Theorem 1.6 taking $q = 5$, we get

$$3T(r, f) \le \sum_{j=1}^{5} \overline{N}(r, a_j) + S(r) \ .$$

Since a_j is completely multiple, we see that for $r \geq 1$,

$$\overline{N}(r, a_j) \leq \frac{1}{2}N(r, a_j) \leq \frac{1}{2}T(r, f) + h, \quad (j = 1, 2, 3, 4, 5)$$

where h is a positive constant. Consequently

$$\frac{1}{2}T(r, f) \leq 5h + S(r)$$

which leads to a contradiction in taking account of the condition $2°$ in Theorem 1.4.

In particular if $f(z)$ is a transcendental entire function, then $f(z)$ has at most two finite completely multiple values. In fact, if $f(z)$ has three finite completely multiple values $a_j(j = 1, 2, 3)$, then in Theorem 1.6, taking $q = 4, a_4 = \infty$, we get

$$2T(r, f) \leq \sum_{j=1}^{3} \overline{N}(r, a_j) + S(r)$$

which also leads to a contradiction.

After more than fifty years later and by following earlier work of C. Chuang, Frank-Weissenborn and C. Osgood, N. Steinmetz have now been able to present a most convincing proof of the following generalized result which was raised by Nevanlinna in 1929.

Theorem. (Nevanlinna's second fundamental theorem for small functions). Let $f(z)$ be a transcendental meromorphic function and $a_1(z)$, $a_2(z), \ldots, a_q(z)$ be distinct $q(\geq 2)$ meromorphic small functions (including constant function) satisfying

$$T(r, a_i(z)) = S(r, f) \quad \text{as} \quad r \to \infty \quad (i = 1, 2, \ldots, q) .$$

Then, for any $\varepsilon > 0$,

$$\sum_{i=1}^{q} m\left(r, \frac{1}{f - a_i}\right) + m(r, f) \leq (2 + \varepsilon)T(r, f) + S(r, f) .$$

Corollary.
$$\sum_{a(z)} \delta(a(z), f) + \delta(\infty, f) \leq 2 ,$$

where the summation is over all deficient functions of f including constants; $a(z)$ is called a deficient function if $T(r, a(z)) = S(r, f)$ and $1 - \overline{\lim_{r \to \infty}} N\left(r, \dfrac{1}{f - a(z)}\right) \Big/ T(r, f) > 0$.

The basic ingredient of the proof on the above result is the success in replacing the lemma of logarithmic derivative (Theorem 1.2) by an estimation of $m(r, P(f)/f^h)$, $P(f)$ a differential polynomial of f and h a positive integer.

Remark. It is natural and interesting to find some non-trivial applications of the generalized results in the studies of fix-points and factorization theory of meromorphic functions.

1.7. SYSTEMS OF MEROMORPHIC FUNCTIONS

In this paragraph our main purpose is to give a complete proof of the following theorem of Borel on systems of entire functions:

Theorem 1.7. Let $f_j(z)(j = 1, 2, \ldots, n)$ and $g_j(z)(j = 1, 2, \ldots, n)$ $(n \geq 2)$ be two systems of entire functions satisfying the following conditions:

1) $\sum_{j=1}^{n} f_j(z) e^{g_j(z)} \equiv 0$.

2) For $1 \leq j \leq n, 1 \leq h, k \leq n, h \neq k$, the order of $f_j(z)$ is less than the order of $e^{g_h(z) - g_k(z)}$: $\rho(f_j) < \rho(e^{g_h - g_k})$. Then $f_j(z) \equiv 0$ $(j = 1, 2, \ldots, n)$.

This theorem has important applications in the theory of fix points and factorization of meromorphic functions. It gives rise to the research of Nevanlinna on systems of meromorphic functions. His main result is the following theorem:

Theorem 1.8. Let $\varphi_j(z)(j = 1, 2, \ldots, n)$ be n linearly independent meromorphic functions satisfying the identity

$$\varphi_1 + \varphi_2 + \ldots + \varphi_n = 1 . \tag{1.110}$$

Then for $1 \leq j \leq n$, we have

$$T(r, \varphi_j) \leq \sum_{k=1}^{n} N\left(r, \frac{1}{\varphi_k}\right) + N(r, \varphi_j) + N(r, D) + S(r) , \tag{1.111}$$

where

$$D = \begin{vmatrix} \varphi_1 & \varphi_2 & \cdots & \varphi_n \\ \varphi_1' & \varphi_2' & \cdots & \varphi_n' \\ . & . & \cdots & . \\ \varphi_1^{(n-1)} & \varphi_2^{(n-1)} & \cdots & \varphi_n^{(n-1)} \end{vmatrix} , \qquad (1.112)$$

and when r is exterior to a sequence of intervals $\{J_p\}$ of finite total length,

$$S(r) = O\{\log T(r) + \log r\} , \qquad (1.113)$$

where

$$T(r) = \max_{1 \leq j \leq n} T(r, \varphi_j) . \qquad (1.114)$$

Proof. Differentiating (1.110) successively, we get

$$\varphi_1^{(k)} + \varphi_2^{(k)} + \ldots + \varphi_n^{(k)} = 0 \quad (k = 1, 2, \ldots, n-1) . \qquad (1.115)$$

Since $\varphi_j (j = 1, 2, \ldots, n)$ are linearly independent, D is not identically equal to zero and, by (1.110) and (1.115), we have

$$D = D_j \quad (j = 1, 2, \ldots, n) , \qquad (1.116)$$

where D_j is the minor corresponding to φ_j in D. Hence

$$\varphi_1 = \frac{D_1}{\varphi_2 \cdots \varphi_n} \bigg| \frac{D}{\varphi_1 \varphi_2 \cdots \varphi_n} = \frac{\Delta_1}{\Delta} , \qquad (1.117)$$

where

$$\Delta = \begin{vmatrix} 1 & 1 & \cdots & 1 \\ \dfrac{\varphi_1'}{\varphi_1} & \dfrac{\varphi_2'}{\varphi_2} & \cdots & \dfrac{\varphi_n'}{\varphi_n} \\ . & . & \cdots & . \\ \dfrac{\varphi_1^{(n-1)}}{\varphi_1} & \dfrac{\varphi_2^{(n-1)}}{\varphi_2} & \cdots & \dfrac{\varphi_n^{(n-1)}}{\varphi_n} \end{vmatrix} \qquad (1.118)$$

and Δ_1 is the minor corresponding to the element 1 in the first row and the first column of Δ. From (1.117), we get

$$m(r, \varphi_1) \leq m(r, \Delta_1) + m\left(r, \frac{1}{\Delta}\right) \leq m(r, \Delta_1) + m(r, \Delta) + N(r, \Delta) + h , \qquad (1.119)$$

where h is a constant. Next from

$$\Delta = \frac{D}{\varphi_1 \varphi_2 \ldots \varphi_n} \, ,$$

we get

$$N(r, \Delta) \le N(r, D) + \sum_{j=1}^{n} N\left(r, \frac{1}{\varphi_j}\right) \, . \tag{1.120}$$

On the other hand, if we set

$$S_1(r) = m(r, \Delta_1) + m(r, \Delta) + h \, ,$$

then by (1.118) and Corollary 1.2, it is easy to see that there is a sequence of intervals $\{J_p\}$ depending only on $\varphi_j (j = 1, 2, \ldots n)$ and of finite total length, such that when r is exterior to $\{J_p\}$, we have

$$S_1(r) = O\{\log T(r) + \log r\} \, . \tag{1.121}$$

Consequently

$$T(r, \varphi_1) \le \sum_{k=1}^{n} N\left(r, \frac{1}{\varphi_k}\right) + N(r, D) + N(r, \varphi_1) + S_1(r) \, . \tag{1.122}$$

Similarly we have

$$T(r, \varphi_j) \le \sum_{k=1}^{n} N\left(r, \frac{1}{\varphi_k}\right) + N(r, D) + N(r, \varphi_j) + S_j(r)$$
$$(j = 2, 3, \ldots, n) \, , \tag{1.123}$$

where $S_j(r)$ is such that, when r is exterior to $\{J_p\}$, we have

$$S_j(r) = O\{\log T(r) + \log r\} \quad (j = 2, 3, \ldots, n) \, . \tag{1.124}$$

Finally defining

$$S(r) = \max_{l \le j \le n} S_j(r)$$

we get (1.111) and (1.113) from (1.122), (1.123), (1.121) and (1.124).

Theorem 1.9. Let $f_j(z)(j = 1, 2, \ldots, n; n \ge 2)$ be n meromorphic functions satisfying the following conditions:

$1°$ $\sum_{j=1}^{n} c_j f_j(z) \equiv 0$, where $c_j (j = 1, 2, \ldots, n)$ are constants.

$2°$ $f_j(z) \not\equiv 0$ $(j = 1, 2, \ldots, n)$ and for $1 \le j, k \le n, j \ne k$, $f_j(z)/f_k(z)$ is not a rational function.

$3°$ $N(r, f_j) = o\{\tau(r)\}$, $N(r, 1/f_j) = o\{(\tau(r)\}(j = 1, 2, \ldots, n)$, where

$$\tau(r) = \min_{\substack{1 \le j,k \le n \\ j \ne k}} \left\{ T\left(r, \frac{f_j}{f_k}\right) \right\} .$$

Then $c_j = 0$ $(j = 1, 2, \ldots, n)$.

Proof. Consider first the case $n = 2$. Then

$$c_1 f_1(z) + c_2 f_2(z) \equiv 0 .$$

Assume that $c_j (j = 1, 2)$ are not both equal to zero, for example $c_1 \ne 0$, then

$$\frac{f_1(z)}{f_2(z)} \equiv -\frac{c_2}{c_1} ,$$

which is incompatible with condition $2°$. Consequently Theorem 1.9 holds when $n = 2$.

Now assume that Theorem 1.9 holds for an integer $n \ge 2$, and let us show that it is also true for $n + 1$. In fact, consider $n + 1$ meromorphic functions $f_j(z)(j = 1, 2, \ldots, n+1)$ satisfying the conditions in Theorem 1.9, so that

$$\sum_{j=1}^{n+1} c_j f_j(z) \equiv 0 . \tag{1.125}$$

Suppose that $c_j (j = 1, 2, \ldots, n + 1)$ are not all equal to zero. Then $c_j (j = 1, 2, \ldots, n + 1)$ must be all different from zero. In fact, if for example $c_{n+1} = 0$, then

$$\sum_{j=1}^{n} c_j f_j(z) \equiv 0$$

and $f_j(z)(j = 1, 2, \ldots, n)$ satisfy the conditions in Theorem 1.9. Since by assumption Theorem 1.9 holds for the integer n, we have $c_j = 0$ $(j = 1, 2, \ldots, n)$ and hence $c_j = 0$ $(j = 1, 2, \ldots, n + 1)$, contrary to the hypothesis that c_j $(j = 1, 2, \ldots, n + 1)$ are not all equal to zero.

So $c_j \ne 0$ $(j = 1, 2, \ldots, n + 1)$. Set

$$\varphi_j(z) = -\frac{c_j f_j(z)}{c_{n+1} f_{n+1}(z)} \quad (j = 1, 2, \ldots, n) . \tag{1.126}$$

By (1.125), we have

$$\sum_{j=1}^{n} \varphi_j(z) \equiv 1 \ .$$

Next we are going to show that $\varphi_j(z)(j = 1, 2, \ldots, n)$ are linearly independent. In fact, if there are constants, a_j $(j = 1, 2, \ldots, n)$ not all equal to zero, such that

$$\sum_{j=1}^{n} a_j \varphi_j(z) \equiv 0 \ ,$$

then

$$\sum_{j=1}^{n} a_j c_j f_j(z) \equiv 0 \ .$$

Since, by assumption, Theorem 1.9 holds for the integer n, we have $a_j c_j = 0$ $(j = 1, 2, \ldots, n)$. On the other hand, at least one of $a_j (j = 1, 2, \ldots, n)$ is different from zero, for instance $a_1 \neq 0$, so $c_1 = 0$ contrary to the result $c_j \neq 0 (j = 1, 2, \ldots, n+1)$ obtained above.

It follows that we can apply Theorem 1.8 to the functions $\varphi_j(z)(j = 1, 2, \ldots, n)$ defined by (1.126), and get

$$T(r, \varphi_j) \leq \sum_{k=1}^{n} N\left(r, \frac{1}{\varphi_k}\right) + N(r, \varphi_j) + N(r, D) + S(r) \tag{1.127}$$

$$(j = 1, 2, \ldots, n) \ .$$

By the condition $3°$ in Theorem 1.9, it is easy to see that $N(r, \varphi_j)$, $N(r, 1/\varphi_j)$ and $N(r, D)$ are all equal to $o\{T(r)\}$ and so

$$T(r, \varphi_j) \leq o\{T(r)\} + S(r) \quad (j = 1, 2, \ldots, n) \ ,$$
$$T(r) \leq o\{T(r)\} + S(r) \ , \tag{1.128}$$

where $S(r)$ satisfies (1.113) when $r \notin \{J_p\}$. But (1.128) is impossible, because by the condition $2°$ for $f_j (j = 1, 2, \ldots, n+1)$ in Theorem 1.9 and by (1.126), $\varphi_j(z)$ is not a rational function and therefore

$$\lim_{r \to \infty} \frac{T(r)}{\log r} = \infty \ .$$

This contradiction proves that the coefficients $c_j(j = 1, 2, \ldots, n+1)$ in (1.125) are all equal to zero, as we want to show.

Theorem 1.10. Let $f_j(z)$ $(j = 1, 2, \ldots, n)$ and $g_j(z)$ $(j = 1, 2, \ldots, n)$ $(n \geq 2)$ be two systems of entire functions satisfying the following conditions:

1) $\sum_{j=1}^{n} f_j(z)e^{g_j(z)} \equiv 0$.
2) For $1 \leq j, k \leq n, j \neq k, g_j(z) - g_k(z)$ is non-constant.
3) For $1 \leq j \leq n, 1 \leq h, k \leq n, h \neq k$,

$$T(r, f_j) = o\{T(r, e^{g_h - g_k})\} . \tag{1.129}$$

Then $f_j(z) \equiv 0$ $(j = 1, 2, \ldots, n)$.

Proof. Consider first the case $n = 2$. Then

$$f_1(z)e^{g_1(z)} + f_2(z)e^{g_2(z)} \equiv 0 .$$

If $f_i(z)(j = 1, 2)$ are not both identically equal to zero, for instance $f_1(z) \not\equiv 0$. Then

$$e^{g_1(z) - g_2(z)} \equiv -\frac{f_2(z)}{f_1(z)} ,$$

and by the condition 3), we get

$$\begin{aligned} T(r, e^{g_1 - g_2}) = T(r, f_2/f_1) &\leq T(r, f_2) + T(r, 1/f_1) \\ &= T(r, f_2) + T(r, f_1) + a = o\{T(r, e^{g_1 - g_2})\} , \end{aligned}$$

which is impossible.

Now assume that Theorem 1.10 holds for an integer $n \geq 2$. To show that it also holds for $n+1$, let $f_j(z), g_j(z)$ $(j = 1, 2, \ldots, n+1)$ satisfy the conditions in Theorem 1.10, and assume that $f_j(z)$ $(j = 1, 2, \ldots, n+1)$ are not all identically equal to zero. Then we must have

$$f_j(z) \not\equiv 0 \quad (j = 1, 2, \ldots, n+1) .$$

In fact, if for instance $f_{n+1}(z) \equiv 0$, then

$$\sum_{j=1}^{n} f_j(z)e^{g_j(z)} \equiv 0$$

and $f_j(z)$ $(j = 1, 2, \ldots, n)$ satisfy the conditions in Theorem 1.10, consequently, by assumption, $f_j(z) \equiv 0$ $(j = 1, 2, \ldots, n)$ and hence $f_j(z) \equiv 0$ $(j = 1, 2, \ldots, n+1)$ contrary to the hypothesis.

Now set

$$F_j(z) = f_j(z)e^{g_j(z)}, \quad c_j = 1 \ (j = 1, 2, \ldots, n+1) \tag{1.130}$$

then by condition 1)

$$\sum_{j=1}^{n+1} c_j F_j(z) \equiv 0 . \tag{1.131}$$

Obviously $F_j(z) \not\equiv 0$ $(j = 1, 2, \ldots, n+1)$ and, Theorem 1.10 being true for $n = 2$, when $1 \le j, k \le n+1, j \ne k$, $F_j(z)/F_k(z)$ is not a rational function. Moreover,

$$N(r, F_j) = 0 \quad (j = 1, 2, \ldots, n+1)$$

and, for $h \ne k$ $(1 \le h, k \le n+1)$ we have

$$N\left(r, \frac{1}{F_j}\right) = N\left(r, \frac{1}{f_j}\right) \le T\left(r, \frac{1}{f_j}\right) = T(r, f_j) + \beta = o\{T(r, e^{g_h - g_k})\} ,$$

$$\frac{F_h}{F_k} = \frac{f_h}{f_k} e^{g_h - g_k} ,$$

$$T(r, e^{g_h - g_k}) = T\left(r, \frac{f_k}{f_h} \frac{F_h}{F_k}\right) \le T(r, f_k) + T(r, f_h) + T\left(r, \frac{F_h}{F_k}\right) + \beta'$$

$$= T\left(r, \frac{F_h}{F_1}\right) + o\{T(r, e^{g_h - g_k})\} ,$$

$$\frac{1}{2} T(r, e^{g_h - g_k}) < T\left(r, \frac{F_h}{F_k}\right) ,$$

$$N\left(r, \frac{1}{F_j}\right) = o\left\{T\left(r, \frac{F_h}{F_k}\right)\right\} .$$

So $F_j(z)$ $(j = 1, 2, \ldots, n+1)$ satisfy the conditions $1°, 2°, 3°$ in Theorem 1.9 and consequently $c_j = 0$ $(j = 1, 2, \ldots, n+1)$ contrary to (1.130). This proves Theorem 1.10.

Finally for the proof of Theorem 1.7, we still need the following two lemmas:

Lemma 1.2. Let

$$g(z) = c_0 z^m + c_1 z^{m-1} + \ldots + c_m \quad (m \ge 1, c_0 \ne 0)$$

be a polynomial and define

$$A(r, g) = \max_{|z|=r} \text{Re}\{g(z)\} .$$

Then when r is sufficiently large, we have

$$\frac{1}{2}|c_0|r^m < A(r, g) < 2|c_0|r^m . \tag{1.132}$$

Proof. Set

$$z = re^{i\theta}, \quad c_j = |c_j|e^{i\alpha_j} \quad (j = 0, 1, 2, \ldots m) .$$

Then

$$g(re^{i\theta}) = |c_0|r^m e^{i(m\theta+\alpha_0)} + |c_1|r^{m-1}e^{i((m-1)\theta+\alpha_1)}$$
$$+ \ldots + |c_m|e^{i\alpha_m} ,$$
$$\text{Re}\{g(re^{i\theta})\} = |c_0|r^m \cos(m\theta + \alpha_0) + |c_1|r^{m-1}\cos((m-1)\theta + \alpha_1)$$
$$+ \ldots + |c_m|\cos\alpha_m . \tag{1.133}$$

In particular taking $\theta = \theta_0 = -\alpha_0/m$, we get

$$\text{Re}\{g(re^{i\theta_0})\} = |c_0|r^m + |c_1|r^{m-1}\cos((m-1)\theta_0 + \alpha_1)$$
$$+ \ldots + |c_m|\cos\alpha_m .$$

Since

$$\text{Re}\{g(re^{i\theta_0})\} \le A(r, g) ,$$

obviously the first part of (1.132) holds. By (1.133), the second part of (1.132) also holds.

Lemma 1.3. Let $f(z)$ be a transcendental entire function and set $F(z) = e^{f(z)}$. Then

$$\lim_{r \to \infty} \frac{\log \log M(r, F)}{\log r} = \infty . \tag{1.134}$$

Proof. We have $M(r, F) = e^{A(r,f)}$ and the inequality

$$M(r, f) \le \frac{2R}{R-r}\{A(R, f) + 2|f(0)|\} \quad (0 < r < R) .$$

(See, for instance, Lemma 1.2 in the book *Singular Directions of Meromorphic Functions*.) In particular, taking $R = 2r$, we get

$$M(r, f) \le 4\{A(2r, f) + 2|f(0)|\} .$$

On the other hand, by (1.39), we have

$$\lim_{r \to \infty} \frac{\log M(r, f)}{\log r} = \infty .$$

Hence

$$\lim_{r \to \infty} \frac{\log A(r, f)}{\log r} = \infty$$

and (1.134) follows.

Now let us prove Theorem 1.7 as follows: By Theorem 1.10, it is sufficient to show that the condition 2) in Theorem 1.7 implies the conditions 2) and 3) in Theorem 1.10. In fact, since the order of an entire function is always non-negative, hence for $h \ne k$, the order of $e^{g_h(z) - g_k(z)}$ is greater than zero and $g_h(z) - g_k(z)$ is non-constant. So the condition 2) in Theorem 1.10 is satisfied. To see the condition 3) in Theorem 1.10 is also satisfied, consider a function $f_j(z)$ and two functions $g_h(z), g_k(z)(h \ne k)$. Since the order of $f_j(z)$ is less than that of $e^{g_h(z) - g_k(z)}$, the order ρ of $f_j(z)$ is finite. Distinguish two cases:

A. $g_h(z) - g_k(z)$ is a polynomial of degree $m \ge 1$. In this case, by Lemma 1.2, evidently the order of $e^{g_h(z) - g_k(z)}$ is m. Hence $\rho < m$. Taking a number λ such that $\rho < \lambda < m$, then when r is sufficiently large,

$$T(r, f_j) < r^\lambda$$

and by Lemma 1.2,

$$T(2r, f_j) = o\{\log M(r, e^{g_h - g_k})\} . \tag{1.135}$$

On the other hand, by (1.33),

$$\log M(r, e^{g_h - g_k}) \le 3T(2r, e^{g_h - g_k}) .$$

Consequently,

$$T(r, f_j) = o\{T(r, e^{g_h - g_k})\} . \tag{1.136}$$

B. $g_h(z) - g_k(z)$ is a transcendental entire function. In this case, by Lemma 1.3,

$$\lim_{r \to \infty} \frac{\log \log M\left(r, e^{g_h - g_k}\right)}{\log r} = \infty .$$

Taking two numbers λ, λ' such that $\rho < \lambda < \lambda'$ when r is sufficiently large, we have

$$\log M\left(r, e^{g_h - g_k}\right) > r^{\lambda'} .$$

It follows that (1.135) and (1.136) are also true. So condition 3) in Theorem 1.10 is also satisfied. Theorem 1.7 is now completely proved.

Corollary 1.7. Let $f_j(z)(j = 1, 2, \ldots, n+1)$ and $g_j(z)(j = 1, 2, \ldots, n)$ $(n \geq 1)$ be two systems of entire functions satisfying the following conditions:

1) $\sum_{j=1}^{n} f_j(z)e^{g_j(z)} \equiv f_{n+1}(z)$.

2) For $1 \leq j \leq n+1, 1 \leq h \leq n$, the order of $f_j(z)$ is less than the order of $e^{g_h(z)}$. In case $n \geq 2$, for $1 \leq j \leq n+1, l \leq h, k \leq n, h \neq k$, the order of $f_j(z)$ is less than the order of $e^{g_h(z) - g_k(z)}$.

Then $f_j(z) \equiv 0$ $(j = 1, 2, \ldots, n+1)$.

Proof. It is sufficient to note that, in setting $g_{n+1}(z) = 0$, we have

$$\sum_{j=1}^{n} f_j(z)e^{g_j(z)} - f_{n+1}(z)e^{g_{n+1}(z)} \equiv 0 ,$$

and then Corollary 1.7 follows from Theorem 1.7.

2

FIX-POINTS OF MEROMORPHIC FUNCTIONS

2.1. INTRODUCTION

In the theory of fix-points of meromorphic functions, Nevanlinna's theory of meromorphic functions and Montel's theory of normal families play an important role. So in this chapter we first give some complements of Nevanlinna's theory of meromorphic functions sketched out in the previous chapter and prove some theorems on fix-points due to Rosenbloom and Baker. Next we give an account of the main points of Montel's theory of normal families of holomorphic functions and apply it to Fatou's theory of fix-points of entire functions.

2.2. SOME THEOREMS ON MEROMORPHIC FUNCTIONS

We first prove the following theorem which is a generalization of Nevanlinna's second fundamental theorem.

Theorem 2.1. Let $f(z)$ be a non-constant meromorphic function. Let $\varphi_j(z)(j = 1, 2, 3)$ be three distinct meromorphic functions such that

$$T(r, \varphi_j) = \mathrm{o}\{T(r, f)\} \quad (j = 1, 2, 3) . \tag{2.1}$$

Then for $r > 1$ we have the inequality

$$T(r, f) \leq \sum_{j=1}^{3} \overline{N}\left(r, \frac{1}{f - \varphi_j}\right) + \mathrm{o}\{T(r, f)\} + S(r) \tag{2.2}$$

where $S(r)$ satisfies the following conditions:

1° If $f(z)$ is of finite order, then

$$S(r) = O(\log r) \ .$$

2° In general, there exists a sequence of intervals $\{I_p\}$ of finite total length such that for $r \notin \{I_p\}$ we have

$$S(r) = O\{\log T(r, f) + \log r\} \ .$$

Proof. Consider the auxiliary function

$$F(z) = \frac{f(z) - \varphi_1(z)}{f(z) - \varphi_3(z)} \frac{\varphi_2(z) - \varphi_3(z)}{\varphi_2(z) - \varphi_1(z)} \ . \tag{2.3}$$

We have

$$T(r, f) \leq T(r, f - \varphi_3) + T(r, \varphi_3) + \log 2 = T\left(r, \frac{1}{f - \varphi_3}\right) + \lambda + T(r, \varphi_3) + \log 2$$

where λ is a constant. Then by the identity

$$1 + \frac{\varphi_3 - \varphi_1}{f - \varphi_3} = \frac{\varphi_2 - \varphi_1}{\varphi_2 - \varphi_3} F \ ,$$

we get

$$T\left(r, \frac{1}{f - \varphi_3}\right) \leq T\left(r, \frac{1}{\varphi_3 - \varphi_1}\right) + T\left(r, \frac{\varphi_3 - \varphi_1}{f - \varphi_3}\right)$$

$$\leq T\left(r, \frac{1}{\varphi_3 - \varphi_1}\right) + T\left(r, \frac{\varphi_3 - \varphi_1}{\varphi_2 - \varphi_3} F\right) + \log 2$$

$$\leq T\left(r, \frac{1}{\varphi_3 - \varphi_1}\right) + T\left(r, \frac{\varphi_2 - \varphi_1}{\varphi_2 - \varphi_3}\right) + T(r, F) + \log 2 \ . \tag{2.4}$$

From (2.3) and (2.4) we deduce

$$T(r, f) \leq 3 \sum_{j=1}^{3} T(r, \varphi_j) + T(r, F) + \lambda_1 \tag{2.5}$$

where λ_1 is a constant. (2.5) shows that the function $F(z)$ is non-constant, consequently by Theorem 1.6, we have, for $r > 0$,

$$T(r, F) \leq \overline{N}\left(r, \frac{1}{F}\right) + \overline{N}\left(r, \frac{1}{F - 1}\right) + \overline{N}(r, F) + S(r) \tag{2.6}$$

where $S(r)$ satisfies the conditions $1°$ and $2°$ in Theorem 1.4 with respect to $F(z)$.

We have

$$\overline{N}(r, F) \leq \overline{N}\left(r, \frac{f - \varphi_1}{f - \varphi_3}\right) + \overline{N}\left(r, \frac{\varphi_2 - \varphi_3}{\varphi_2 - \varphi_1}\right)$$

$$= \overline{N}\left(r, 1 + \frac{\varphi_3 - \varphi_1}{f - \varphi_3}\right) + \overline{N}\left(r, \frac{\varphi_2 - \varphi_3}{\varphi_2 - \varphi_1}\right)$$

$$\leq \overline{N}\left(r, \frac{1}{f - \varphi_3}\right) + \overline{N}(r, \varphi_3 - \varphi_1) + \overline{N}\left(r, \frac{\varphi_2 - \varphi_3}{\varphi_2 - \varphi_1}\right) .$$

Since

$$\overline{N}(r, \varphi_3 - \varphi_1) \leq T(r, \varphi_3 - \varphi_1), \quad \overline{N}\left(r, \frac{\varphi_2 - \varphi_3}{\varphi_2 - \varphi_1}\right) \leq T\left(r, \frac{\varphi_2 - \varphi_3}{\varphi_2 - \varphi_1}\right) ,$$

we get

$$\overline{N}(r, F) \leq \overline{N}\left(r, \frac{1}{f - \varphi_3}\right) + o\{T(r, f)\} . \tag{2.7}$$

Similarly we get

$$\overline{N}\left(r, \frac{1}{F}\right) \leq \overline{N}\left(r, \frac{1}{f - \varphi_1}\right) + o\{T(r, f)\} , \tag{2.8}$$

$$\overline{N}\left(r, \frac{1}{F - 1}\right) \leq \overline{N}\left(r, \frac{1}{f - \varphi_2}\right) + o\{T(r, f)\} . \tag{2.9}$$

Inequalities (2.5)-(2.9) yield

$$T(r, f) \leq \sum_{j=1}^{3} \overline{N}\left(r, \frac{1}{f - \varphi_j}\right) + o\{T(r, f)\} + S(r) .$$

It remains to show that $S(r)$ satisfies the conditions $1°$ and $2°$ in Theorem 2.1. In fact we have

$$T(r, F) \leq T\left(r, \frac{f - \varphi_1}{f - \varphi_3}\right) + T\left(r, \frac{\varphi_2 - \varphi_3}{\varphi_2 - \varphi_1}\right)$$

and by the identity

$$\frac{f - \varphi_1}{f - \varphi_3} = 1 + \frac{\varphi_3 - \varphi_1}{f - \varphi_3} ,$$

we have

$$T(r, F) \leq T(r, f) + 3 \sum_{j=1}^{3} T(r, \varphi_j) + \lambda_2$$

where λ_2 is a constant. Now it is evident that $S(r)$ also satisfies the conditions 1° and 2° in Theorem 2.1.

Similarly we can prove the following theorem:

Theorem 2.2. Let $f(z)$ be a non-constant meromorphic function. Let $\varphi_j(z)(j = 1, 2)$ be two distinct meromorphic functions such that

$$T(r, \varphi_j) = \mathrm{o}\{T(r, f)\} \quad (j = 1, 2) .$$

Then for $r > 1$ we have the inequality

$$T(r, f) \leq \overline{N}(r, f) + \sum_{j=1}^{2} \overline{N}\left(r, \frac{1}{f - \varphi_j}\right) + \mathrm{o}\{T(r, f)\} + S(r)$$

where $S(r)$ satisfies the conditions 1° and 2° in Theorem 2.1.

Now we are going to prove some theorems on the growth of composite functions. For this we need some preliminary lemmas and theorems.

Lemma 2.1. Let a and b be two positive numbers such that $b \geq 8a^2$. Then for $x \geq 2$, we have

$$e^{\frac{b}{a}} x > 8a \log x + 8b . \tag{2.10}$$

Proof. Consider the auxiliary function

$$\varphi(x) = e^{\frac{b}{a}} x - 8a \log x - 8b .$$

Evidently it is sufficient to show that

$$\varphi(2) > 0, \quad \varphi'(x) = e^{\frac{b}{a}} - \frac{8a}{x} > 0 \quad (x \geq 2) . \tag{2.11}$$

We have

$$2e^{\frac{b}{a}} > 2\left(1 + \frac{b}{a} + \frac{1}{2}\frac{b^2}{a^2}\right) > \left(1 + \frac{b}{a}\right)^2 , \tag{2.12}$$

$$1 + \frac{b}{a} > 8a ,$$

$$\left(1 + \frac{b}{a}\right)^2 > 8a\left(1 + \frac{b}{a}\right) = 8a + 8b \tag{2.13}$$

$$> 8a \log 2 + 8b$$

and hence

$$2e^{\frac{b}{a}} > 8a \log 2 + 8b$$

which shows that the first inequality in (2.11) holds. On the other hand, (2.12) and (2.13) imply

$$2e^{\frac{b}{a}} > 8a \ ,$$

hence the second inequality in (2.11) also holds.

Lemma 2.2. Let $U(r)$ be a non-negative and non-decreasing function in an interval $0 < r < \rho$. Let a and b be two positive numbers such that $b \geq 2a$ and $b \geq 8a^2$. Assume that the inequality

$$U(r) < a \log^+ U(R) + a \log \frac{R}{R-r} + b \tag{2.14}$$

holds for $0 < r < R < \rho$. Then the inequality

$$U(r) < 2a \log \frac{R}{R-r} + 2b$$

holds for $0 < r < R < \rho$.

This Lemma in a different form was obtained by Borel in his fundamental paper "Sur les zéros des fonctions entières", *Acta Math.* **20** (1897). It is put in the present form by Bureau and Milloux.

Proof. Let r and R be two values such that $0 < r < R < \rho$ and assume

$$U(r) \geq 2a \log \frac{R}{R-r} + 2b \ . \tag{2.15}$$

We are going to show that the two values $r' = (r + R)/2$ and R also satisfy the inequality

$$U(r') \geq 2a \log \frac{R}{R-r'} + 2b \ . \tag{2.16}$$

In fact, by (2.14), we have

$$U(r) < a \log^+ U(r') + a \log \frac{r'}{r'-r} + b$$
$$< a \log^+ U(r') + a \log \frac{R}{R-r'} + b \ .$$

From this inequality and (2.15) written in the form

$$U(r) \geq 2a \log \frac{R}{R-r'} - 2a \log 2 + 2b \ ,$$

we get

$$\log^+ U(r') > \log \frac{R}{R - r'} - 2\log 2 + \frac{b}{a} \, ,$$

that is

$$\log^+ U(r') > \log \left(\frac{1}{4} e^{\frac{b}{a}} \frac{R}{R - r'} \right) \, .$$

Since $b \geq 2a$, we have

$$\frac{1}{4} e^{\frac{b}{a}} \frac{R}{R - r'} > \frac{1}{4} e^{\frac{b}{a}} \geq \frac{1}{4} e^2 > 1 \, ,$$

$$U(r') > \frac{1}{4} e^{\frac{b}{a}} \frac{R}{R - r'} \, . \tag{2.17}$$

On the other hand, by Lemma 2.1, taking $x = R/(R - r')$ in (2.10), we have

$$e^{\frac{b}{a}} \frac{R}{R - r'} > 8a \log \frac{R}{R - r'} + 8b \, . \tag{2.18}$$

Inequalities (2.17) and (2.18) yield (2.16).

Now let

$$r_n = (r_{n-1} + R)/2 \quad (n = 1, 2, \ldots), \quad r_0 = r \, .$$

Then for each n, we have

$$0 < r_n < R < \rho \tag{2.19}$$

and

$$U(r_n) \geq 2a \log \frac{R}{R - r_n} + 2b \, . \tag{2.20}$$

$U(r)$ being a non-decreasing function, (2.19) implies $U(r_n) \leq U(R)$. On the other hand, (2.20) implies $U(r_n) \to \infty$, as $n \to \infty$. So we get a contradiction.

Lemma 2.3. Let $a > e$ and $x > 0$ be two numbers. Then we have

$$\log x + a \log^+ \log^+ \frac{1}{x} \leq a(\log a - 1) + \log^+ x \, . \tag{2.21}$$

Proof. If $x > 1/e$, then $\log^+ \log^+ (1/x) = 0$, hence (2.21) holds. If $x \leq 1/e$, then (2.21) becomes

$$\log x + a \log \log \frac{1}{x} \leq a(\log a - 1) \, . \tag{2.22}$$

Consider the function

$$\varphi(y) = a \log y - y - a(\log a - 1) \quad (y > 0) \tag{2.23}$$

and its derivative

$$\varphi'(y) = \frac{a}{y} - 1 . \tag{2.24}$$

We can see easily

$$\varphi(y) \le \varphi(a) = 0 \quad (y > 0) . \tag{2.25}$$

Replacing in (2.25) y by $\log(1/x)$, we get (2.22).

Theorem 2.3. Let $f(z)$ be a holomorphic function in the circle $|z| < 1$ in which $f(z)$ does not take the values 0 and 1. Then for $|z| < 1$, we have

$$\log |f(z)| < \frac{1}{1 - |z|} \left(A \log^+ |f(0)| + B \log \frac{2}{1 - |z|} \right) \tag{2.26}$$

where A and B are two positive numerical constants.

This is the classical theorem of Schottky.

Proof. Consider the auxiliary function

$$F(z) = \frac{1}{f(z)} + \frac{1}{f(z) - 1} .$$

As in the proof of Theorem 1.4, we see that for $0 < r < 1$, we have

$$m(r, F) \ge m \left(r, \frac{1}{f} \right) + m \left(r, \frac{1}{f - 1} \right) - 2 \log 4 - \log 3$$

and

$$m(r, F) \le m(r, f) + 2m \left(r, \frac{f'}{f} \right) + m \left(r, \frac{f'}{f - 1} \right) + \log \frac{1}{|f'(0)|} + \log 2$$

where we assume $f'(0) \ne 0$. Hence

$$m \left(r, \frac{1}{f} \right) + m \left(r, \frac{1}{f - 1} \right) \le m(r, f) + 2m \left(r, \frac{f'}{f} \right)$$
$$+ m \left(r, \frac{f'}{f - 1} \right) + \log \frac{1}{|f'(0)|} + \alpha$$

where α is a positive numerical constant. Then by (1.14), we have

$$m(r, f) = m\left(r, \frac{1}{f}\right) + \log|f(0)| \, ,$$

$$m(r, f - 1) = m\left(r, \frac{1}{f-1}\right) + \log|f(0) - 1|$$

and get

$$m(r, f) \le 2m\left(r, \frac{f'}{f}\right) + m\left(r, \frac{f'}{f-1}\right) + \log|f(0)| + \log|f(0) - 1|$$

$$+ \log\frac{1}{|f'(0)|} + \alpha' \, .$$

Next by Theorem 1.2, for $0 < r < R < 1$, we have

$$m\left(r, \frac{f'}{f}\right) < 4\log^+ m(R, f) + 3\log\frac{1}{R-r}$$

$$+ 2\log\frac{1}{r} + 4\log^+\log^+\frac{1}{|f(0)|} + 16 \, ,$$

$$m\left(r, \frac{f'}{f-1}\right) < 4\log^+ m(R, f) + 3\log\frac{1}{R-r}$$

$$+ 2\log\frac{1}{r} + 4\log^+\log^+\frac{1}{|f(0) - 1|} + 16 \, .$$

Hence

$$m(r, f) < 12\log^+ m(R, f) + 9\log\frac{1}{R-r}$$

$$+ 6\log\frac{1}{r} + 8\log^+\log^+\frac{1}{|f(0)|} + \log|f(0)|$$

$$+ 4\log^+\log^+\frac{1}{|f(0) - 1|}$$

$$+ \log|f(0) - 1| + \log\frac{1}{|f'(0)|} + \alpha'' \, .$$

$$(2.27)$$

By Lemma 2.3,

$$8\log^+\log^+\frac{1}{|f(0)|} + \log|f(0)| \le 8(\log 8 - 1) + \log^+|f(0)| \, , \quad (2.28)$$

$$4\log^+\log^+\frac{1}{|f(0) - 1|} + \log|f(0) - 1| \le 4(\log 4 - 1) + \log^+|f(0)| + \log 2 \, .$$

$$(2.29)$$

From (2.27), (2.28) and (2.29), we get the following result: For $1/2 \leq r < R < 1$, we have

$$m(r, f) < 12 \log^+ m(R, f) + 9 \log \frac{R}{R-1} + 2 \log^+ |f(0)| + \log \frac{1}{|f'(0)|} + \beta$$

where β is a positive numerical constant. In the interval $0 < r < 1$ define a function $U(r)$ as follows:

$$U(r) = 0 \quad \left(0 < r < \frac{1}{2}\right), \quad U(r) = m(r, f) \quad \left(\frac{1}{2} \leq r < 1\right).$$

Then $U(r)$ is non-negative and non-decreasing for $0 < r < 1$, and satisfies, for $0 < r < R < 1$, the inequality

$$U(r) < a \log^+ U(R) + a \log \frac{R}{R-r} + b$$

where

$$a = 12, \quad b = 2 \log^+ |f(0)| + \log^+ \frac{1}{|f'(0)|} + \beta'$$

with $\beta' = \max(\beta, 8a^2)$. Consequently by Lemma 2.2, for $0 < r < R < 1$, we have

$$U(r) < 2a \log \frac{R}{R-r} + 2b .$$

Keeping r fixed and letting $R \to 1$, we get

$$U(r) \leq 2a \log \frac{1}{1-r} + 2b .$$

Hence for $1/2 \leq r < 1$,

$$m(r, f) \leq 2a \log \frac{1}{1-r} + 2b$$

and then for $0 < r < 1$,

$$m(r, f) < 2a \log \frac{1}{1-r} + 2b'$$

with $b' = a \log 2 + b$. By (1.33), in taking $R = 1, \rho = (1+r)/2$, we have

$$\log^+ M(r, f) \leq \frac{4}{1-r} m \left(\frac{1+r}{2}, f\right) \quad (0 < r < 1) .$$

Hence for $0 < r < 1$, we have

$$\log^+ M(r, f) < \frac{4}{1-r} \left(2a \log \frac{2}{1-r} + 4 \log^+ |f(0)| + 2 \log^+ \frac{1}{|f'(0)|} + b_1 \right)$$

$$(2.30)$$

where b_1 is a positive numerical constant.

Now let us distinguish two cases:

$1°$ $|f'(0)| \geq 1$. In this case, by (2.30), we have for $0 < r < 1$,

$$\log M(r, f) < \frac{4}{1-r} \left(2a \log \frac{2}{1-r} + 4 \log^+ |f(0)| + b_1 \right) . \qquad (2.31)$$

$2°$ $|f'(0)| < 1$. In this case, consider a value $0 < r < 1$ and let $z_0 = re^{i\theta_0}$ be a point of the circle $|z| = r$ such that

$$|f(z_0)| = M(r, f) .$$

Consider the segment $S : z = te^{i\theta_0} (0 \leq t \leq r)$ and distinguish two cases:

1) On S we always have $|f'(z)| \leq 1$. Then

$$|f(z_0)| \leq |f(0)| + r$$

and

$$\log M(r, f) \leq \log^+ |f(0)| + \log 2 . \qquad (2.32)$$

2) On S we do not always have $|f'(z)| \leq 1$. Then we can get a point $z_1 = t_1 e^{i\theta_0} (0 < t_1 < r)$ of S, such that on the segment $S_1 : z = te^{i\theta_0}$ $(0 \leq t < t_1)$ we have $|f'(z)| < 1$ and $|f'(z_1)| = 1$. Consider the function

$$F(\varsigma) = f\{z_1 + \varsigma(1 - |z_1|)\}$$

which is holomorphic in the circle $|\varsigma| < 1$ and does not take the values 0 and 1. Hence by (2.30), in the circle $|\varsigma| < 1$ we have

$$\log |F(\varsigma)| < \frac{4}{1-|\varsigma|} \left(2a \log \frac{2}{1-|\varsigma|} + 4 \log^+ |F(0)| + 2 \log^+ \frac{1}{|F'(0)|} + b_1 \right) .$$

$$(2.33)$$

We have

$$|F(0)| = |f(z_1)| \leq |f(0)| + t_1 , \quad |F'(0)| = |f'(z_1)|(1 - |z_1|) = 1 - |z_1| .$$

In (2.33), taking in particular

$$\varsigma = \frac{z_0 - z_1}{1 - |z_1|} = \frac{(r - t_1)e^{i\theta_0}}{1 - t_1}$$

we get

$$\log M(r, f) < \frac{4}{1 - r} \left(2a \log \frac{2}{1 - r} + 4 \log^+ |f(0)| \right.$$

$$\left. + 2 \log \frac{1}{1 - r} + b_1 + \log 2 \right) . \qquad (2.34)$$

Consequently in the two cases, by (2.31), (2.32) and (2.34) we have for $0 < r < 1$,

$$\log M(r, f) < \frac{1}{1 - r} \left(16 \log^+ |f(0)| + B \log \frac{2}{1 - r} \right)$$

where B is a positive numerical constant. Theorem 2.3 is therefore proved.

Theorem 2.4. Let $w = f(z)$ be a holomorphic function in the circle $|z| < 1$ satisfying the condition:

$$f(0) = 0 , \quad M\left(\frac{1}{2}, f\right) \geq 1 .$$

Then in the circle $|z| < 1$ the function $f(z)$ takes every value w on a certain circle $|w| = r$, with $r > A$, A being a positive numerical constant.

This is a theorem of Bohr, H. (Über einen Satz von Edmund Landau, *Scripta Univ. Hierosolymitanarum*, **1** (1923)).

Proof. Consider a number $R > 0$ and assume that for each number $r \geq R$, there exists a point w_0 of the circle $|w| = r$, such that the function $f(z)$ does not take the value w_0 in the circle $|z| < 1$. Then there are two points w_1 and w_2 respectively on the circles $|w| = R$ and $|w| = 2R$, such that $f(z)$ does not take the values w_1 and w_2 in the circle $|z| < 1$. Consider the function

$$g(z) = \frac{f(z) - w_1}{w_2 - w_1}$$

which is holomorphic in the circle $|z| < 1$ and does not take the values 0 and 1, and we have

$$|g(0)| = \left| \frac{w_1}{w_2 - w_1} \right| \leq 1 .$$

Consequently by Theorem 2.3, for $|z| < 1$ we have

$$|g(z)| < \exp\left(\frac{B}{1-|z|}\log\frac{2}{1-|z|}\right)$$

and

$$|f(z)| \leq |w_1| + |w_2 - w_1||g(z)| \leq R(1 + 3|g(z)|) .$$

Let z_0 be a point of the circle $|z| = 1/2$ such that

$$|f(z_0)| = M\left(\frac{1}{2}, f\right) ,$$

then we get

$$1 < kR, \quad k = 1 + 3e^{2B\log 4} .$$

Hence there is a circle $|w| = r, r \geq 1/k$, such that in the circle $|z| < 1, f(z)$ takes every value w of the circle $|w| = r$.

Now we are going to prove some theorems on the growth of composite functions.

Theorem 2.5. Let $g(z)$ and $h(z)$ be two entire functions and let $f(z) = g\{h(z)\}$. If $h(0) = 0, h(z) \not\equiv 0$, then there is a number c $(0 < c < 1)$ such that for $r > 0$, we have

$$M(r, f) \geq M\left\{cM\left(\frac{r}{2}, h\right), g\right\} . \tag{2.35}$$

This theorem is due to Pólya, G. ("On an integral function of an integral function", *J. London Math. Soc.*, **1** (1926) 12-15).

Proof. Consider the function

$$H(Z) = \frac{h(rZ)}{M\left(\frac{r}{2}, h\right)}$$

which evidently satisfies the conditions of Theorem 2.4. There is then a circle $|w| = R, R \geq c$ $(0 < c < 1)$, such that in the circle $|Z| < 1$, the function $H(Z)$ takes every value w of the circle $|w| = R$. Hence in the circle $|z| < r$, the function $h(z)$ takes every value W of the circle $|W| = RM\left(\frac{r}{2}, h\right)$. Set

$$R' = RM\left(\frac{r}{2}, h\right) .$$

Let W_0 be a point of the circle $|W| = R'$, such that $|g(W_0)| = M(R',g)$ and let z_0 be a point of the circle $|z| < r$, such that $h(z_0) = W_0$. Then we have

$$M(r,f) \geq |f(z_0)| = |g\{h(z_0)\}| = |g(W_0)|$$
$$= M(R',g) \geq M\left\{cM\left(\frac{r}{2},h\right),g\right\} .$$

Corollary 2.1. Let $g(z)$ and $h(z)$ be two entire functions and let $f(z) = g\{h(z)\}$. If $h(z)$ is non-constant, there is a number c_1 $(0 < c_1 < 1)$ such that when r is sufficiently large, we have

$$M(r,f) \geq M\left\{c_1 M\left(\frac{r}{2},h\right),g\right\} . \tag{2.36}$$

Proof. Let $g_1(z) = g(z + h(0)), h_1(z) = h(z) - h(0)$. Then $f(z) = g_1\{h_1(z)\}$ and by Theorem 2.5,

$$M(r,f) \geq M\left\{cM\left(\frac{r}{2},h_1\right),g_1\right\} .$$

Since

$$g(z) = g_1(z - h(0)), \quad h(z) = h_1(z) + h(0) ,$$

we have

$$M(r,g) \leq M(r + |h(0)|,g_1) , \quad M(r,h) \leq M(r,h_1) + |h(0)| .$$

Hence when r is sufficiently large, we have

$$M(r,g) \leq M(2r,g_1), \quad M(r,h) < 2M(r,h_1)$$

and

$$M(r,f) \geq M\left\{\frac{c}{2}M\left(\frac{r}{2},h\right),g_1\right\} \geq M\left\{\frac{c}{4}M\left(\frac{r}{2},h\right),g\right\} .$$

Now we prove another theorem of Pólya, also based upon Theorem 2.4.

Theorem 2.6. Let $f(z)$ and $g(z)$ be two non-constant entire functions such that the order of the function $\varphi(z) = g\{f(z)\}$ is finite. Then either $f(z)$ is a polynomial or the order of $g(z)$ is zero.

Proof. Assume that $f(z)$ and $g(z)$ are both transcendental entire functions. We are going to show that the order of $g(z)$ is zero. Let $r > 0$ and $c = f(0)$. Then the function

$$F(Z) = \frac{f\left(\frac{1}{2}rZ\right) - c}{M\left(\frac{1}{4}r, f - c\right)}$$

satisfies the condition of Theorem 2.4. Hence there is a circle $|w| = R$ $(R > A)$ such that in the circle $|Z| < 1$ the function $F(Z)$ takes every value w of the circle $|w| = R$. It follows that in the circle $|z| < r/2$ the function $f(z)$ takes every value W of the circle $|W - c| = RM\left(\frac{r}{4}, f - c\right)$. Set

$$R' = RM\left(\frac{r}{4}, f - c\right) > AM\left(\frac{r}{4}, f - c\right) \geq A\left\{M\left(\frac{r}{4}, f\right) - |c|\right\}$$

and let W_0 be a point of the circle $|W - c| = R'$ such that

$$|g(W_0)| = \max_{|W - c| = R'} |g(W)| \, ,$$

and let z_0 be a point of the circle $|z| < r/2$ such that

$$f(z_0) = W_0 \, .$$

Then

$$M\left(\frac{r}{2}, \varphi\right) \geq |\varphi(z_0)| = |g\{f(z_0)\}| = |g(W_0)|$$

$$= \max_{|W - c| = R'} |g(W)| \geq \max_{|W| = R' - |c|} |g(W)| = M(R' - |c|, g)$$

where r is assumed to be sufficiently large such that $R' > |c|$. We have

$$R' - |c| > AM\left(\frac{r}{4}, f\right) - |c|(A + 1) \, .$$

Since $f(z)$ is a transcendental entire function, for any positive integer N, we have

$$AM\left(\frac{r}{4}, f\right) > r^N + |c|(A + 1)$$

provided that r is sufficiently large. Hence

$$R' - |c| > r^N \, , \quad M\left(\frac{r}{2}, \varphi\right) \geq M(r^N, g) \, . \tag{2.37}$$

By hypothesis, the order of $\varphi(z)$ is finite, hence

$$\log M(r, \varphi) < r^k$$

when r is sufficiently large, where $k > 0$ is a constant. So we have, for sufficiently large t,

$$\log M(t, g) < t^{\frac{k}{N}} \, .$$

Since N may be taken arbitrarily large, the order of $g(z)$ is zero.

Theorem 2.7. Let $f(z)$ and $g(z)$ be two transcendental entire functions and let $\varphi(z) = g\{f(z)\}$. Then

$$\lim_{r \to \infty} \frac{T(r, \varphi)}{T(r, g)} = \infty . \qquad (2.38)$$

Proof. In the proof of Theorem 2.6, we have obtained (2.37), where N is any positive integer, provided that r is sufficiently large. We have

$$\log M\left(\frac{r}{2}, \varphi\right) \le \frac{r + \frac{r}{2}}{r - \frac{r}{2}} T(r, \varphi) = 3T(r, \varphi) ,$$

$$T(r^N, g) \le \log M(r^N, g) ,$$

hence

$$T(r^N, g) \le 3T(r, \varphi) .$$

Since $T(r, g)$ is a convex function of $\log r$ we see that the function

$$\frac{T(r, g) - T(r_0, g)}{\log r - \log r_0}$$

is non-decreasing for $r > r_0$, where $r_0 > 1$. Hence

$$\frac{T(r^N, g) - T(r_0, g)}{N \log r - \log r_0} \ge \frac{T(r, g) - T(r_0, g)}{\log r - \log r_0}$$

and, when r is sufficiently large, we have

$$T(r^N, g) \ge \frac{N}{2}\{T(r, g) - T(r_0, g)\} > \frac{N}{4}T(r, g) ,$$

and

$$T(r, \varphi) > \frac{N}{12}T(r, g) .$$

Since N is arbitrary, we have (2.38).

Theorem 2.8. Let $f(z)$ be a transcendental entire function and $g(z)$ a transcendental meromorphic function. Let $\varphi(z) = g\{f(z)\}$. Then

$$\lim_{r \to \infty} \frac{T(r, \varphi)}{T(r, f)} = \infty . \qquad (2.39)$$

Proof. Assume first that $g(z)$ has an infinite number of zeros. Choose p zeros $a_i (i = 1, 2, \ldots, p)$ of $g(z)$, such that

$$|a_i - a_j| > 1 \quad (i, j = 1, 2, \ldots, p; i \neq j) \ .$$

Then it is easily seen that we can find a number $0 < \delta \leq 1/2$ and a number $M > 0$ such that in the p circles $c_i : |z - a_i| < \delta (i = 1, 2, \ldots, p)$ we have respectively the p inequalities

$$|g(z)| < M|z - a_i| \quad (i = 1, 2, \ldots, p) \ .$$

Evidently any point z belongs at most to one of the circles $c_i (i = 1, 2, \ldots, p)$. Consequently

$$\log^+ \frac{1}{|g(z)|} \geq \sum_{i=1}^{p} \log^+ \frac{\delta}{|z - a_i|} - \log^+ \delta M \ ,$$

$$\log^+ \frac{1}{|\varphi(z)|} \geq \sum_{i=1}^{p} \log^+ \frac{\delta}{|f(z) - a_i|} - \log^+ \delta M \ ,$$

and

$$m\left(r, \frac{1}{\varphi}\right) \geq \sum_{i=1}^{p} m\left(r, \frac{\delta}{f - a_i}\right) - \log^+ \delta M \ .$$

Then making use of the inequality

$$m\left(r, \frac{1}{f - a_i}\right) \leq m\left(r, \frac{\delta}{f - a_i}\right) + \log \frac{1}{\delta} \ ,$$

we get

$$\sum_{i=1}^{p} m\left(r, \frac{1}{f - a_i}\right) \leq m\left(r, \frac{1}{\varphi}\right) + h \tag{2.40}$$

where h is a constant. On the other hand, we see easily that

$$\sum_{i=1}^{p} N\left(r, \frac{1}{f - a_i}\right) \leq N\left(r, \frac{1}{\varphi}\right) \ . \tag{2.41}$$

Inequalities (2.40) and (2.41) yield

$$\sum_{i=1}^{p} T\left(r, \frac{1}{f - a_i}\right) \leq T\left(r, \frac{1}{\varphi}\right) + h \ .$$

Finally making use of the relations

$$T\left(r, \frac{1}{f - a_i}\right) = T(r, f) + O(1), \quad T\left(r, \frac{1}{\varphi}\right) = T(r, \varphi) + \lambda ,$$

we get

$$T(r, \varphi) \geq pT(r, f) + O(1) ,$$

and when r is sufficiently large,

$$\frac{T(r, \varphi)}{T(r, f)} > \frac{p}{2} .$$

Since p is arbitrary, we have (2.39).

In the above we have assumed that $g(z)$ has an infinite number of zeros. In general, since $g(z)$ is a transcendental meromorphic function, there is a finite value w_0 such that the function $g_1(z) = g(z) - w_0$ has an infinite number of zeros. Set

$$\varphi_1(z) = g_1\{f(z)\} = g\{f(z)\} - w_0 = \varphi(z) - w_0 .$$

By the result just obtained, we have

$$\lim_{r \to \infty} \frac{T(r, \varphi_1)}{T(r, f)} = \infty .$$

Then by the inequality

$$T(r, \varphi_1) \leq T(r, \varphi) + \log^+ |w_0| + \log 2 ,$$

we get again (2.39).

Note that in the proof of Theorem 2.8, the condition that $f(z)$ is a transcendental entire function is not necessary. What is important is that $f(z)$ is a non-constant entire function.

2.3. SOME THEOREMS OF ROSENBLOOM ON FIX-POINTS

For reference, see "The fix points of entire functions", *Medd. Lunds Univ. Mat. Sem., Suppl.-Bd. M. Riesz* **186** (1952).

Definition 2.1. Let $f(z)$ be a meromorphic function. A point $z_0 (z_0 \neq \infty)$ is said to be a fix-point of $f(z)$, if $f(z_0) = z_0$. This is equivalent to say that z_0 is a zero of the function $f(z) - z$.

Theorem 2.9. Let $P(z)$ be a polynomial of degree $n \geq 2$ and let $f(z)$ be a transcendental entire function. Then the function $P\{f(z)\}$ has an infinite number of fix-points.

Proof. Suppose, on the contrary, that the function $P\{f(z)\}$ has at most a finite number of fix points. Then the function $P\{f(z)\} - z$ has at most a finite number of zeros, hence

$$P\{f(z)\} - z = Q(z)e^{\alpha(z)} \tag{2.42}$$

where $Q(z) \not\equiv 0$ is a polynomial and $\alpha(z)$ is an entire function. Since the left hand member of (2.42) is a transcendental entire function, so $\alpha(z)$ is non-constant.

The equation $f\{P(z)\} = z$ has necessarily at most a finite number of solutions. In fact, if z_0 is a solution of this equation, then

$$f\{P(z_0)\} = z_0 \ ,$$

hence

$$P(f\{P(z_0)\}) = P(z_0)$$

and by (2.42),

$$Q\{P(z_0)\} = 0 \ .$$

So z_0 is a zero of the polynomail $Q\{P(z)\}$. Since $Q\{P(z)\} \not\equiv 0$, it has at most a finite number of zeros.

By Theorem 2.8,

$$\lim_{r \to \infty} \frac{T(r, f(P))}{T(r, P)} = \infty \ ,$$

hence $f\{P(z)\}$ is a transcendental entire function and we have

$$f\{P(z)\} = L(z)e^{\beta(z)} + z \tag{2.43}$$

where $L(z) \not\equiv 0$ is a polynomial and $\beta(z)$ is a non-constant entire function.

From (2.42) and (2.43) we get

$$P\{L(z)e^{\beta(z)} + z\} = P(f\{P(z)\}) = Q\{P(z)\}e^{\alpha\{P(z)\}} + P(z) \ . \tag{2.44}$$

Let

$$P(z) = c_0 z^n + c_1 z^{n-1} + \ldots + c_n \quad (c_0 \neq 0, \quad n \geq 2) \ .$$

After calculation, we get

$$\sum_{j=1}^{n} u_j(z)e^{j\beta(z)} + v(z)e^{\alpha\{P(z)\}} = 0 \tag{2.45}$$

where $u_j(z)(j = 1, 2, \ldots, n)$ and $v(z) = -Q\{P(z)\} \not\equiv 0$ are polynomials. It is easy to see that the polynomial $u_j(z)$ and the polynomial

$$c_0 \binom{n}{j} z^{n-j} \{L(z)\}^j \qquad (2.46)$$

have the same degree and have the same coefficient for the term of the highest degree. Consequently the polynomials $u_j(z)$ $(j = 1, 2, \ldots, n)$ and $v(z)$ are all non-identically equal to zero.

Now distinguish two cases:

1) The functions $j\beta(z) - \alpha\{P(z)\}(j = 1, 2, \ldots, n)$ are all non-constant. In this case, by (2.45) and Theorem 1.7, $u_j(z)(j = 1, 2, \ldots, n)$ and $v(z)$ are all identically equal to zero. So we get a contradiction.

2) There is an integer j_0 $(1 \leq j_0 \leq n)$ such that $j_0\beta(z) - \alpha\{P(z)\}$ is a constant. In this case, for $j \neq j_0$ $(1 \leq j \leq n), j\beta(z) - \alpha\{P(z)\}$ is non-constant, for otherwise, $\beta(z)$ would be a constant. Writing (2.45) in the form

$$\sum_{j=1}^{n}{}' u_j(z)e^{j\beta(z)} + v_1(z)e^{\alpha\{P(z)\}} = 0$$

where in the sum $\sum'$ the term $u_j(z)e^{j_0\beta(z)}$ is omitted. Since $n \geq 2$ and $j\beta(z) - \alpha\{P(z)\}(1 \leq j \leq n, j \neq j_0)$ are non-constant, we get again a contradiction by Theorem 1.7.

Lemma 2.4. Let $\varphi(z)$ be a non-constant entire function. Then there do not exist two pairs of functions $f_i(z), g_i(z)(i = 1, 2)$ satisfying the following conditions:

1) $f_i(z)(i = 1, 2)$ are two transcendental entire functions each of which has at most a finite number of fix-points.

2) $g_i(z)(i = 1, 2)$ are two distinct entire functions.

3) The identities $\varphi(z) = f_i\{g_i(z)\}$ $(i = 1, 2)$ hold.

Lemma 2.5. Let $\varphi(z)$ be a non-constant meromorphic function. Then there do not exist three pairs of functions $f_i(z), g_i(z)(i = 1, 2, 3)$ satisfying the following conditions:

1) $f_i(z)(i = 1, 2, 3)$ are three transcendental meromorphic functions each of which has at most a finite number of fix-points.

2) $g_i(z)(i = 1, 2, 3)$ are three distinct entire functions.

3) The identities $\varphi(z) = f_i\{g_i(z)\}(i = 1, 2, 3)$ hold.

We first prove Lemma 2.5 as follows. Assume on the contrary, that there exist $f_i(z), g_i(z)(i = 1, 2, 3)$ satisfying the conditions 1), 2), 3). Then the identities

$$\varphi(z) = f_i\{g_i(z)\} \quad (i = 1, 2, 3)$$

imply that $g_i(z)(i = 1, 2, 3)$ are non-constant. Consequently by Theorem 2.8, we have

$$T(r, g_i) = o\{T(r, \varphi)\} \quad (i = 1, 2, 3) . \tag{2.47}$$

Then by Theorem 2.1, for $r > 1$, we have

$$T(r, \varphi) \leq \sum_{i=1}^{3} \overline{N}\left(r, \frac{1}{\varphi - g_i}\right) + o\{T(r, \varphi)\} + S(r) \tag{2.48}$$

where $S(r)$ satisfies the conditions $1°$ and $2°$ in Theorem 1.4 with respect to $\varphi(z)$. Since by hypothesis $f_i(z)$ has at most a finite number of fix-points $z_{ik}(k = 1, 2, \ldots, p_i)$, we have

$$\overline{N}\left(r, \frac{1}{\varphi - g_i}\right) = \overline{N}\left(r, \frac{1}{f_i(g_i) - g_i}\right) \leq \sum_{k=1}^{p_i} \overline{N}\left(r, \frac{1}{g_i - z_{ik}}\right) .$$

From (2.47), we have

$$\overline{N}\left(r, \frac{1}{g_i - z_{ik}}\right) = o\{T(r, \varphi)\} \quad (k = 1, 2, \ldots, p_i) ,$$

hence

$$\overline{N}\left(r, \frac{1}{\varphi - g_i}\right) = o\{T(r, \varphi)\} \quad (i = 1, 2, 3) .$$

Then by (2.48) we get

$$T(r, \varphi) \leq o\{T(r, \varphi)\} + S(r) . \tag{2.49}$$

Since $g_i(z)$ is non-constant, (2.47) implies

$$\lim_{r \to \infty} \frac{T(r, \varphi)}{\log r} = \infty . \tag{2.50}$$

Next since $S(r)$ satisfies the conditions $1°$ and $2°$ in Theorem 2.1 with respect to $\varphi(z)$, we get a contradiction from (2.49) and (2.50).

By the same method we can prove Lemma 2.4 base upon Theorem 2.2.

Theorem 2.10. Let $F(z)$ and $G(z)$ be two transcendental entire functions. Then at least one of the two functions $F(z)$ and $F\{G(z)\}$ has an infinite number of fix-points.

Proof. Assume, on the contrary, that the conclusion of this theorem is untrue. Consider the functions

$$\varphi(z) = F\{G(z)\} \, ,$$
$$f_1(z) = F(z), \quad f_2(z) = F\{G(z)\}, \quad g_1(z) = G(z), \quad g_2(z) = z \, .$$

By Theorem 2.8, we have

$$\lim_{r \to \infty} \frac{T(r, \varphi)}{T(r, G)} = \infty \, ,$$

hence $\varphi(z)$ is a transcendental entire function. We have

$$\varphi(z) = f_1\{g_1(z)\}, \quad \varphi(z) = f_2\{g_2(z)\} \, .$$

So by the assumption, the conditions 1), 2), 3) in Lemma 2.4 are all satisfied, but this is impossible.

Corollary 2.2. Let $f(z)$ be a transcendental entire function. Then the function $f\{f(z)\}$ has an infinite number of fix-points.

Proof. By Theorem 2.10, at least one of the two functions $f(z)$ and $f\{f(z)\}$ has an infinite number of fix-points. Noting that if z_0 is a fix-point of $f(z)$, then $f(z_0) = z_0, f\{f(z_0)\} = f(z_0) = z_0$, and hence z_0 is also a fix-point of $f\{f(z)\}$. It follows that $f\{f(z)\}$ has always an infinite number of fix-points.

Theorem 2.11. Let $F(z)$ be a transcendental meromorphic function and $G(z)$ and $H(z)$ two transcendental entire functions. Then at least one of the three functions $F(z), F\{G(z)\}$ and $F\{G[H(z)]\}$ has an infinite number of fix-points.

Assume, on the contrary, that the conclusion of this theorem is untrue. Consider the functions

$$\varphi(z) = F\{G[H(z)]\} \, ,$$

$$f_1(z) = F(z), \qquad f_2(z) = F\{G(z)\}, \quad f_3(z) = F\{G[H(z)]\} ;$$
$$g_1(z) = G[H(z)], \quad g_2(z) = H(z), \qquad g_3(z) = z .$$

By Theorem 2.8, we have

$$\lim_{r \to \infty} \frac{T(r, g_1)}{T(r, g_2)} = \infty ,$$

hence the functions $g_i(z)(i = 1, 2, 3)$ are distinct and $g_1(z)$ is a transcendental entire function. Also by Theorem 2.8, we see that $f_2(z)$ and $f_3(z)$ are transcendental meromorphic functions. On the other hand, evidently the identities

$$\varphi(z) = f_i\{g_i(z)\} \quad (i = 1, 2, 3)$$

hold. Hence $f_i(z), g_i(z)$ $(i = 1, 2, 3)$ satisfy the conditions 1), 2), 3) in Lemma 2.5 and we get a contradiction.

2.4. SOME THEOREMS OF BAKER ON FIX-POINTS

For refernce, see "The existence of fix-points of entire functions", *Math. Zeit* **73** (1960).

Definition 2.2. Let $f(z)$ be an entire function. Set

$$f_1(z) = f(z), \quad f_2(z) = f\{f_1(z)\}, \quad f_3(z) = f\{f_2(z)\}, \ldots ,$$

in general

$$f_{n+1}(z) = f\{f_n(z)\} \quad (n = 1, 2, \ldots) .$$

Let $n \geq 2$ be an integer. Then a fix-point of $f_n(z)$ is called a fix-point of order n of $f(z)$; if z_0 is a fix-point of order n of $f(z)$, but of no lower order, then z_0 is called a fix-point of exact order n of $f(z)$. A fix-point of order 1 of $f(z)$ is also called a fix-point of exact order 1 of $f(z)$.

Theorem 2.12. Let $f(z)$ be a transcendental entire function. Then for each positive integer n, $f(z)$ has an infinite number of fix-points of exact order n, except at most for one exceptional positive integer n.

Proof. Note first that if $n = h + k$, where h and k are two positive integers, then

$$f_n(z) = f_k\{f_h(z)\}$$

and, by Theorem 2.8, we have

$$\lim_{r \to \infty} \frac{T(r, f_n)}{T(r, f_h)} = \infty .$$

Now suppose that there is a positive integer n such that $f(z)$ has at most a finite number of fix-points $z_j (j = 1, 2, \ldots, p)$ of exact order n. Consider a positive integer $m > n$. If z_0 is a root of the equation

$$f_m(z) = f_{m-n}(z) ,$$

then

$$f_n\{f_{m-n}(z_0)\} = f_{m-n}(z_0)$$

and hence $Z = f_{m-n}(z_0)$ is a fix-point of $f_n(z)$. It follows that, either for a certain $j (1 \leq j \leq p)$ we have

$$Z = f_{m-n}(z_0) = z_j ;$$

or Z is a fix-point of exact order $n_0 (1 \leq n_0 \leq n - 1)$ of $f(z)$, in the latter case, we have

$$f_{n_0}(Z) = Z ,$$

that is

$$f_{m-n+n_0}(z_0) = f_{m-n}(z_0) .$$

Summing up these results, we conclude that z_0 is a zero of one of the following functions:

$$f_{m-n}(z) - z_j \quad (j = 1, 2, \ldots, p) ,$$
$$f_{m-n+n_0}(z) - f_{m-n}(z) \quad (n_0 = 1, 2, \ldots, n - 1)$$

and hence

$$\overline{N}\left(r, \frac{1}{f_m - f_{m-n}}\right) \leq \sum_{j=1}^{p} \overline{N}\left(r, \frac{1}{f_{m-n} - z_j}\right)$$
$$+ \sum_{n_0=1}^{n-1} \overline{N}\left(r, \frac{1}{f_{m-n+n_0} - f_{m-n}}\right)$$
$$= O\left\{\sum_{i=1}^{m-1} T(r, f_i)\right\} = o\{T(r, f_m)\} .$$

$$(2.51)$$

Now applying Theorem 2.2, with $f_m(z)$ for f and $f_{m-n}(z), z$ for φ_j $(j = 1, 2)$, we get, for $r > 1$,

$$T(r, f_m) \leq \overline{N}\left(r, \frac{1}{f_m - f_{m-n}}\right) + \overline{N}\left(r, \frac{1}{f_m - z}\right) + \mathrm{o}\{T(r, f_m)\} + S(r)$$

where $S(r)$ satisfies the conditions 1° and 2° in Theorem 2.1 with respect to $f_m(z)$. Consequently when r is exterior to a sequence of intervals of finite total length, we have

$$T(r, f_m) \leq \overline{N}\left(r, \frac{1}{f_m - z}\right) + \mathrm{o}\{T(r, f_m)\} . \tag{2.52}$$

Now divide the zeros of $f_m(z) - z$ into two kinds:

1) A zero z_0 of $f_m(z) - z$ belongs to the first kind, if z_0 is not a zero of one of the functions $f_k(z) - z$ $(k = 1, 2, \ldots, m - 1)$, in other words, z_0 is a fix-point of exact order m of $f(z)$.

2) A zero z_0 of $f_m(z) - z$ belongs to the second kind, if z_0 is a zero of one of the functions $f_k(z) - z$ $(k = 1, 2, \ldots, m - 1)$. By (2.52) and the relation

$$\sum_{k=1}^{m-1} \overline{N}\left(r, \frac{1}{f_k - z}\right) = \mathrm{O}\left\{\sum_{k=1}^{m-1} T(r, f_k)\right\} = \mathrm{o}\{T(r, f_m)\} ,$$

evidently $f_m(z) - z$ has an infinite number of zeros of the first kind, hence $f(z)$ has an infinite number of fix-points of exact order m.

So far we have proved that if there is a positive integer n such that $f(z)$ has at most a finite number of fix-points of exact order n, then for any positive integer $m > n$, $f(z)$ has an infinite number of fix-points of exact order m. The conclusion of Theorem 2.12 is therefore valid.

Theorem 2.13. Let $f(z)$ be a polynomial of degree $d \geq 2$. Then for each positive integer n, $f(z)$ has at least one fix-point of exact order n, except at most one positive integer n.

Proof. Evidently the degree of $f_n(z)$ is d^n and hence for each positive integer n, $f(z)$ has fix-point of order n. In particular, $f(z)$ has fix-point of order 1, and hence fix-point of exact order 1.

Suppose, on the contrary, there exist two positive integers n and k such that $n > k \geq 2$ and that $f(z)$ has neither fix-point of exact order n, nor fix-point of exact order k. Consider the rational function

$$\varphi(z) = \frac{f_n(z) - z}{f_{n-k}(z) - z} .$$

Denoting by $\overline{N}_0$ the number of the zeros of $\varphi(z)$, each zero being counted only once, we are going to prove that we have

$$\overline{N}_0 \leq d^{n-2} . \tag{2.53}$$

In fact, if z_0 is a zero of $\varphi(z)$, then z_0 is a zero of $f_n(z) - z$. Since by assumption, $f(z)$ has no fix-point of exact order n, z_0 is a fix-point of exact order $j (1 \leq j < n)$ of $f(z)$. n must be divisible by j. In fact, this is clear, if $j = 1$. If $1 < j < n$ and n is not divisible by j, then $mj < n < (m+1)j$ where m is a positive integer, hence $n = mj + h$ $(1 \leq h < j)$ and

$$f_n(z_0) = f_{mj+h}(z_0) = f_h\{f_{mj}(z_0)\} .$$

Then from

$$f_{2j}(z_0) = f_j\{f_j(z_0)\} = f_j(z_0) = z_0 ,$$
$$f_{3j}(z_0) = f_j\{f_{2j}(z_0)\} = f_j(z_0) = z_0 ,$$
$$\cdots$$
$$f_{mj}(z_0) = z_0 ,$$

we have

$$f_n(z_0) = f_h(z_0) \neq z_0$$

and get a contradiction. Consequently n is divisible by j. Distinguish two cases:

1) $n = 3$. In this case, $j = 1$, and the zeros of $\varphi(z)$ are fix-points of $f(z)$, whose number is at most equal to d, hence (2.53) holds.

2) $n = 4$. In this case, $j = 1$ or $j = 2$. Since a fix-point of order 1 of $f(z)$ is also a fix-point of order 2 of $f(z)$, it follows that the zeros of $\varphi(z)$ are fix-points of $f_2(z)$, whose number is at most d^2, it follows that (2.53) also holds.

3) $n > 4$. In this case, first note that n is not divisible by $n - 1$ and $n - 2$. In fact, if n is divisible by $n - 1$, then $n = m(n-1)$ (m is a positive integer with $1 < m < n$), and therefore $(m - 1)n = m$ which is obviously impossible. Similary, if n is divisible by $n-2$, then $n = m(n-2)(1 < m < n)$ and therefore $(m - 1)n = 2m$, $2n = 2m(n - 2) = (m - 1)n(n - 2)$, $2 = (m - 1)(n - 2)$ which is impossible. Hence $j \leq n - 3$ and each zero of $\varphi(z)$

must be a fix-point of one of the polynomials $f_j(z)(j = 1, 2, \ldots, n-3)$. Consequently

$$\overline{N}_0 \leq \sum_{j=1}^{n-3} d^j = \frac{d^{n-2} - d}{d-1} \leq d^{n-2} - d < d^{n-2} .$$

(2.53) therefore always holds.

Now denoting by $\overline{N}_1$ the number of zeros of $\varphi(z) - 1$, each zero being counted only once, we are going to prove that

$$\overline{N}_1 \leq d^{n-1} . \tag{2.54}$$

In fact, if z_0 is a zero of $\varphi(z) - 1$, then

$$f_{n-k}(z_0) = f_n(z_0) = f_k\{f_{n-k}(z_0)\}$$

which shows that $Z = f_{n-k}(z_0)$ is a fix-point of $f_k(z)$. Since, by assumption, $f(z)$ has no fix-point of exact order k, hence Z is a fix-point of exact order j with $1 \leq j < k$. As in the above, we see that k is divisible by j. Distinguish two cases:

1) $k \geq 3$. In this case, we see, as in the above, that k is not divisible by $k-1$, hence $1 \leq j \leq k-2$. Since

$$f_j(Z) = Z, \quad Z = f_{n-k}(z_0)$$

shows that z_0 is a zero of the polynomial $f_{n-k+j}(z) - f_{n-k}(z)$, we have

$$\overline{N}_1 \leq \sum_{j=1}^{k-2} d^{n-k+j} < \sum_{j=1}^{n-2} d^j < d^{n-1} .$$

Hence (2.54) holds.

2) $k = 2$. In this case, $j = 1$, and hence z_0 is a zero of the polynomial $f_{n-1}(z) - f_{n-2}(z)$. Consequently $\overline{N}_1 \leq d^{n-1}$ and (2.54) also holds.

Now denoting respectively by N_0, N_1 and N' the number of zeros of $\varphi(z), \varphi(z) - 1$ and $\varphi'(z)$, each zero is counted according to its order of multiplicity. Then we have

$$N_0 + N_1 \leq \overline{N}_0 + \overline{N}_1 + N' ,$$

this is because a zero of order $m > 1$ of $\varphi(z)$ or $\varphi(z) - 1$, is a zero of order $m - 1$ of $\varphi'(z)$. Hence, by (2.53) and (2.54),

$$N_0 + N_1 \leq d^{n-2} + d^{n-1} + N' . \tag{2.55}$$

On the other hand, $\varphi(z)$ has a pole of order $d^n - d^{n-k}$ at infinity. Let p be the number of the other poles of $\varphi(z)$, each pole is counted according to its order of multiplicity. Then the number P of the poles of $\varphi(z)$ (with due count of order of multiplicity) in the extended complex plane is given by

$$P = d^n - d^{n-k} + p . \tag{2.56}$$

Evidently the number of the poles of $\varphi(z) - 1$ in the extended complex plane is also P. For $\varphi'(z)$, it has a pole of order $d^n - d^{n-k} - 1$ at infinity and the number of its other poles (with due count of order of multiplicity) does not exceed $2p$, hence in the extended complex plane, the number P' of the poles of $\varphi'(z)$ satisfies the inequality

$$P' \leq d^n - d^{n-k} - 1 + 2p . \tag{2.57}$$

Now it is well known that in the extended complex plane, the number of the zeros and the number of the poles of a rational function are equal (with due count of order of multiplicity), hence from (2.56) we have

$$N_0 + N_1 = 2P = 2\left(d^n - d^{n-k} + p\right) .$$

On the other hand, from (2.57) we have

$$N' = P' \leq d^n - d^{n-k} - 1 + 2p .$$

Finally from (2.55) we get

$$2(d^n - d^{n-k} + p) \leq d^{n-2} + d^{n-1} + N'$$
$$\leq d^{n-2} + d^{n-1} + d^n - d^{n-k} - 1 + 2p ,$$
$$d^n - d^{n-k} \leq d^{n-2} + d^{n-1} - 1 ,$$
$$d^n \leq d^{n-2} + d^{n-k} + d^{n-1} - 1 \leq 2d^{n-2} + d^{n-1} - 1$$
$$\leq d^{n-1} + d^{n-1} - 1 = 2d^{n-1} - 1 \leq d^n - 1 ,$$

and arrive at a contradiction.

Consider for example the entire function $f(z) = e^z + z$ which has no fix-point, hence the positive integer $n = 1$, is exceptional in the sense of Theorem 2.12. On the other hand, consider the polynomial $f(z) = z^2 - z$. We have $f_2(z) - z = z^3(z - 2)$, hence $f_2(z)$ has two fix-points $z = 0$ and $z = 2$. Since these two points are also fix-points of $f(z)$, hence the positive integer $n = 2$ is exceptional in the sense of Theorem 2.13. An interesting question is that, in the case of transcendental entire functions, whether the exceptional positive integer n in Theorem 2.12 may be different from 1.

2.5. NORMAL FAMILIES OF HOLOMORPHIC FUNCTIONS

In the theory of normal families of holomorphic functions, the following theorem of Montel is fundamental.

Theorem 2.14. Let $f_n(z)(n = 1, 2, \ldots)$ be a sequence of holomorphic functions in a domain D. If this sequence of functions is locally uniformly bounded in D, then we can extract from this sequence of functions a subsequence $f_{n_k}(z)(k = 1, 2, \ldots)$ which converges locally uniformly in D.

To say that the sequence $f_n(z)(n = 1, 2, \ldots)$ is locally uniformly bounded in D means that for each point z_0 in D, there is a circle $c : |z - z_0| < r$ belonging to D and a positive number M such that

$$|f_n(z)| \leq M \quad \text{for } n \geq 1, \ z \in c \ . \tag{2.58}$$

Similarly we define the notion of local uniform convergence in D.

For the proof of this theorem, we need two lemmas.

Lemma 2.6. Let $f_n(z)(n = 1, 2, \ldots)$ be a sequence of functions holomorphic in a circle $|z| \leq R$. If the sequence $f_n(z)(n = 1, 2, \ldots)$ is uniformly bounded for $|z| \leq R$ and converges at each point of a set s for which the point $z = 0$ is a point of accumulation, then the sequence $f_n(z)(n = 1, 2, \ldots)$ converges uniformly in any circle $|z| \leq r$ $(0 < r < R)$.

To say that a function is holomorphic in a circle $|z| \leq R$ means that it is holomorphic in a circle $|z| < \rho$ with $\rho > R$.

Proof. Let

$$f_n(z) = \sum_{k=0}^{\infty} a_k^{(n)} z^k \ .$$

Since the sequence $f_n(z)(n = 1, 2, \ldots)$ is uniformly bounded for $|z| \leq R$, we have

$$|f_n(z) - f_n(0)| \leq |f_n(z)| + |f_n(0)| \leq 2M \quad \text{for } n \geq 1, \quad |z| \leq R$$

where $M > 0$ is a constant independent of n and z. Hence by Schwarz lemma, we have

$$|f_n(z) - f_n(0)| \le \frac{2M}{R}|z| \quad (|z| \le R) .$$

Consider a point z_0 of the circle $|z| \le R$. We have

$$|f_n(0) - f_m(0)| \le |f_n(0) - f_n(z_0)| + |f_n(z_0) - f_m(z_0)| + |f_m(z_0) - f_m(0)|$$
$$\le \frac{4M}{R}|z_0| + |f_n(z_0) - f_m(z_0)| .$$

Given arbitarily a number $\varepsilon > 0$, let z_0 be a point of the set s such that

$$\frac{4M}{R}|z_0| < \frac{\varepsilon}{2} ,$$

and next let N be a positive integer such that

$$|f_n(z_0) - f_m(z_0)| < \frac{\varepsilon}{2} \quad (n \ge N, m \ge N) .$$

Then we have

$$|f_n(0) - f_m(0)| < \varepsilon \quad (n \ge N, m \ge N)$$

which shows that the sequence $a_0^{(n)} = f_n(0)(n = 1, 2, \ldots)$ converges to a limit a_0.

Consider the sequence

$$g_n(z) = \frac{f_n(z) - a_0^{(n)}}{z} = a_1^{(n)} + a_2^{(n)}z + \ldots \quad (n = 1, 2, \ldots) .$$

Since on the circle $|z| = R$ we have

$$|g_n(z)| \le \frac{2M}{R} ,$$

this inequality also holds in the circle $|z| < R$, by the maximum modulus principle. Besides this sequence also converges at each point of the set s. Consequently as in the above, we see that the sequence $a_1^{(n)} = g_n(0)$ $(n = 1, 2, \ldots)$ converges to a limit a_1. Continuing in this way, we see that for any k, the sequence $a_k^{(n)}(n = 1, 2, \ldots)$ converges to a limit a_k.

By Cauchy inequality,

$$|a_k^{(n)}| \le \frac{M}{R^k} , \qquad (2.59)$$

hence

$$|a_k| \le \frac{M}{R^k} , \qquad (2.60)$$

and the series

$$f(z) = \sum_{k=0}^{\infty} a_k z^k$$

converges absolutely in the circle $|z| < R$ and defines a holomorphic function $f(z)$. Set

$$f_n(z) = \sum_{k=0}^{K} a_k^{(n)} z^k + \rho_n(z) , \quad f(z) = \sum_{k=0}^{K} a_k z^k + \rho(z)$$

where

$$\rho_n(z) = \sum_{k=K+1}^{\infty} a_k^{(n)} z^k , \quad \rho(z) = \sum_{k=K+1}^{\infty} a_k z^k .$$

Consider a circle $|z| \le r$ $(0 < r < R)$. By (2.59) and (2.60), we see that in this circle the inequalities

$$|\rho_n(z)| \le M \frac{\theta^{K+1}}{1-\theta} , \quad |\rho(z)| \le M \frac{\theta^{K+1}}{1-\theta}$$

hold, where $\theta = r/R$. Given arbitrarily a number of $\varepsilon > 0$, let K be a positive integer such that

$$M \frac{\theta^{K+1}}{1-\theta} < \frac{\varepsilon}{3}$$

and then let N be a positive integer such that for $n \ge N$, we have

$$\left| \sum_{k=0}^{K} a_k^{(n)} z^k - \sum_{k=0}^{K} a_k z^k \right| < \frac{\varepsilon}{3} \quad (|z| \le r) .$$

For $n \ge N$, evidently

$$|f_n(z) - f(z)| < \varepsilon \quad (|z| \le r)$$

and hence in the circle $|z| \leq r$, the sequence $f_n(z)(n = 1, 2, \ldots)$ converges uniformly to $f(z)$.

Lemma 2.7. Let $f_n(z)(n = 1, 2, \ldots)$ be a sequence of holomorphic functions in a domain D. If this sequence is locally uniformly bounded in D and converges at each point of a set s which has a point of accumulation in D, then this sequence converges locally uniformly in D.

This is a theorem of Vitali.

Proof. Let z_0 be a point of accumulation of the set s in D. Consider any point Z of D such that $Z \neq z_0$. Join z_0 and Z by a polygonal line L in D. By Heine-Borel theorem, we see easily that, under the conditions of Lemma 2.7, we can find a number $d > 0$ having the following property: For any point ς of L, the circle $|z - \varsigma| \leq d$ is interior to D and the sequence $f_n(z)(n = 1, 2, \ldots)$ is uniformly bounded in this circle. Now in the sense from z_0 to Z, take on L a finite number of points $\varsigma_j (j = 0, 1, 2, \ldots, n)$ such that

$$|\varsigma_j - \varsigma_{j-1}| < d/2 \quad (j = 1, 2, \ldots, n), \quad \varsigma_0 = z_0, \quad \varsigma_n = Z.$$

Consider first the point ς_0 and the circle $\overline{\Gamma}_0 : |z - \varsigma_0| \leq d$. Since the sequence $f_n(z)(n = 1, 2, \ldots)$ is uniformly bounded in Γ_0 and converges at each point of the set s for which ς_0 is a point of accumulation, hence by Lemma 2.6, the sequence $f_n(z)(n = 1, 2, \ldots)$ converges uniformly in the circle $\overline{\gamma}_0 : |z - \varsigma_0| \leq d/2$. Next consider the point ς_1 and the circle $\overline{\Gamma}_1 : |z - \varsigma_1| \leq d$. Since the point ς_1 is interior to the circle $\gamma_0 : |z - \varsigma_0| < d/2$, we see again, by Lemma 2.6, that the sequence $f_n(z)(n = 1, 2, \ldots)$ converges uniformly in the circle $\overline{\gamma}_1 : |z - \varsigma_1| \leq d/2$. Continuing in this way, we see finally that the sequence $f_n(z)(n = 1, 2, \ldots)$ converges uniformly in the circle $\overline{\gamma}_n : |z - \varsigma_n| \leq d/2$. Thus we have shown that the sequence $f_n(z)(n = 1, 2, \ldots)$ converges locally uniformly in D.

Now let us return to the proof of Theorem 2.14. First let $z_k (k = 1, 2, \ldots)$ be a sequence of points in D, which has a point of accumulation in D. Since the sequence $f_n(z_1)(n = 1, 2, \ldots)$ is bounded, we can extract from it a convergent subsequence

$$f_{\alpha_1}(z_1), \quad f_{\alpha_2}(z_1), \ldots.$$

In other words, the sequence

$$S_1 : f_{\alpha_1}(z), f_{\alpha_2}(z), \ldots$$

converges at the point z_1. In the same way we can extract from the sequence S_1 a subsequence

$$S_2 : f_{\beta_1}(z),\ f_{\beta_2}(z), \ldots$$

which converges at the point z_2. Then from the sequence S_2 we extract a subsequence

$$S_3 : f_{\gamma_1}(z),\ f_{\gamma_2}(z), \ldots$$

which converges at the point z_3. In general, from the sequence S_{k-1} we extract a subsequence $S_k = f_{\lambda_1}(z), f_{\lambda_2}(z), \ldots$ which converges at the point z_k. Continuing this process we get a sequence of sequences $S_k(k = 1, 2, \ldots)$. Consider the sequence

$$S' : f_{\alpha_1}(z),\ f_{\beta_2}(z),\ f_{\gamma_3}(z), \ldots, f_{\lambda_k}(z), \ldots .$$

The first term, second term and the third term of S' are respectively the first term of S_1, the second term of S_2 and the third term of S_3; in general the kth term of S' is the kth term of S_k. Obviously for each k, beginning from the kth term of S' all the subsequent terms of S' belong to S_k. Consequently the sequence S' converges at each of the points $z_k(k = 1, 2, \ldots)$. On the other hand the sequence S' is locally uniformly bounded in D, by the conditions of Theorem 2.14. Hence by Lemma 2.7, the sequence S' converges locally uniformly in D.

Definition 2.3. Let $\{f(z)\}$ be a family of holomorphic functions in a domain D. If from every sequence $f_n(z)(n = 1, 2, \ldots)$ of this family we can extract a subsequence $f_{n_k}(z)(k = 1, 2, \ldots)$ satisfying one of the following two conditions:

1) $f_{n_k}(z)(k = 1, 2, \ldots)$ is locally uniformly convergent in D;
2) As $k \to \infty$, $f_{n_k}(z)$ tends locally uniformly to infinity in D;

then we say that the family $\{f(z)\}$ is normal in D.

In this definition, the word "family" means "set".

We say that the family $\{f(z)\}$ is normal at a point z_0 of D, if there is a circle $c : |z - z_0| < r$ interior to D such that the family $\{f(z)\}$ is normal in c.

Evidently if the family $\{f(z)\}$ is normal in D, then it is normal at each point of D. Conversely we have the following theorem:

Theorem 2.15. Let $\{f(z)\}$ be a family of holomorphic functions in a domain D. If this family is normal at each point of D, then it is normal in D.

For the proof of this theorem we need the following lemma:

Lemma 2.8. Let $\{f(z)\}$ be a family of holomorphic functions in a domain D and assume that $\{f(z)\}$ is normal at each point of D. Let $S : f_n(z)(n = 1, 2, \ldots)$ be a sequence of the family $\{f(z)\}$ and z_0 and z_0' two points of D. If the sequence S is uniformly convergent in a circle c with center z_0, then there is a subsequence S' of S, such that S' is uniformly convergent in a circle c' with center z_0'.

Proof. Join the two points z_0 and z_0' by a polygonal line L in D. By the conditions of Lemma 2.8 and Heine-Borel theorem, it is easy to see that we can find a number $d > 0$ such that for any point ς of L, the circle $|z - \varsigma| < d$ is interior to D and in this circle the family $\{f(z)\}$ is normal. Now in the sense from z_0 to z_0' take on L a finite number of points $\varsigma_j (j = 0, 1, 2, \ldots, n)$ such that

$$|\varsigma_j - \varsigma_{j-1}| < d \quad (j = 1, 2, \ldots, n), \quad \varsigma_0 = z_0, \quad \varsigma_n = z_0' \ .$$

Consider first the circle $c_0 : |z - \varsigma_0| < d$ and the sequence S. Since the family $\{f(z)\}$ is normal in c_0, we can, by the conditions of Lemma 2.8, extract from S a subsequence S_0 which is locally uniformly convergent in c_0. Next consider the circle $c_1 : |z - \varsigma_1| < d$ and the sequence S_0. Since the point ς_1 is interior to c_0, we see that we can extract from S_0 a subsequence S_1 which is locally uniformly convergent in c_1. Continuing in this way we get finally a sequence S_n which is locally uniformly convergent in the circle $c_n : |z - \varsigma_n| < d$. Evidently this sequence S_n has the required properties of the sequence S' in Lemma 2.8.

Now let us return to the proof of Theorem 2.15. First of all, we can find a sequence of points $z_j (j = 1, 2, \ldots)$ of D such that each point of D is a limiting point of this sequence. Such a sequence of points of D can be obtained in different ways of which the simplest is to take the set of rational points (points of the form $z = r_1 + ir_2, r_1, r_2$ being rational numbers) of D. We know that this set is countable.

Consider a point z_j. By hypothesis there is a circle $c : |z - z_j| < r$ interior to D such that the family $\{f(z)\}$ is normal in c. Let R_j be the least upper bound of the set of such positive numbers r. According to R_j is finite or infinite, denote by c_j respectively the circle $|z - z_j| < R_j/2$ or the circle $|z - z_j| < 1$. The circle c_j so defined is interior to D and the family $\{f(z)\}$ is normal in c_j.

Now let $S : f_n(z)(n = 1, 2, \dots)$ be a sequence of the family $\{f(z)\}$. From S we can extract a subsequence S_1 which satisfies one of the two conditions in Definition 2.3 in c_1. Then from S_1 we can extract a subsequence S_2 which satisfies one of the two conditions in Definition 2.3 in c_2. Continuing in this way and finally taking the diagonal sequence, as in the proof of Theorem 2.14, we get a subsequence S' of the sequence S such that in each circle c_j, S' satisfies one of the two conditions in Definition 2.3. We are going to show that the sequence S' satisfies in D one of the two conditions in Definition 2.3. Distinguish two cases:

$1°$ S' is locally uniformly convergent in c_1. In this case, by Lemma 2.8, S' is locally uniformly convergent in each circle c_j. Consider a point z_0 of D. By hypothesis, the family $\{f(z)\}$ is normal in a circle $\Gamma : |z - z_0| < \rho$ $(0 < \rho < 1)$ interior to D. Let z_j be such that $|z_j - z_0| < \rho'(\rho' < \rho/4)$. Then the circle $|z - z_j| < 2\rho'$ is interior to Γ, and hence the family $\{f(z)\}$ is normal in this circle. Consequently, if R_j is finite, then

$$2\rho' \le R_j, \quad \rho' \le \frac{R_j}{2},$$

and hence the circle $\gamma : |z - z_j| < \rho'$ is interior to the circle c_j. If R_j is infinite, then since $\rho' < 1, \gamma$ is also interior to c_j. So the sequence S' is locally uniformly convergent in the circle γ. Since z_0 is a point in the circle γ, hence the sequence S' is uniformly convergent in a circle with center z_0.

$2°$ As $k \to \infty, S'$ converges locally uniformly to infinity in c_1. In this case, by Lemma 2.8, in each circle c_j, S' converges locally uniformly to infinity as $k \to \infty$. Then, as in the above it can be proved that for each point z_0 of D, there is a circle with center z_0, in which S' converges uniformly to infinity as $k \to \infty$.

Theorem 2.16. Let $\{f(z)\}$ be a family of holomorphic functions in a domain D. Let a and b $(a \ne b)$ be two finite values. If each function $f(z)$ of the family $\{f(z)\}$ does not take the values a and b in D, then the family $\{f(z)\}$ is normal in D.

This is an important theorem of Montel (*Leçons sur les Familles Normales de Fonctions Analytiques et leurs Applications*, Paris, 1927).

Proof. In view of Theorem 2.15, we need only to consider the case that D is a circle $c : |z - z_0| < R$. Evidently it is sufficient to show that from any sequence $f_n(z)(n = 1, 2, \dots)$ of the family $\{f(z)\}$ we can extract a subsequence $f_{n_k}(z)(k = 1, 2, \dots)$ satisfying one of the following two conditions:

1) $f_{n_k}(z)(k = 1, 2, \dots)$ is uniformly convergent in the circle c' : $|z-z_0| < R/4$.

2) As $k \to \infty$, $f_{n_k}(z)$ tends uniformly to infinity in c'. To see this, consider the sequence $f_n(z_0)(n = 1, 2, \dots)$ and distinguish two cases:

$1°$ The sequence $f_n(z_0)(n = 1, 2, \dots)$ is bounded:

$$|f_n(z_0)| \leq M \quad (n = 1, 2, \dots) .$$

Applying Theorem 2.3 to the function

$$F(Z) = \frac{f_n(z_0 + RZ) - a}{b - a} \quad (|Z| < 1) ,$$

we get

$$\log |f_n(z) - a| < \frac{|b - a|R}{R - |z - z_0|}$$
$$\times \left(A \log^+ \frac{|f_n(z_0) - a|}{|b - a|} + B \log \frac{2R}{R - |z - z_0|} \right) . \quad (2.61)$$

In particular for $|z - z_0| < R/2$, we have

$$\log |f_n(z) - a| < 2|b - a| \left(A \log^+ \frac{M + |a|}{|b - a|} + B \log 4 \right) .$$

Hence the sequence $f_n(z)(n = 1, 2, \dots)$ is uniformly bounded in the circle $|z - z_0| < R/2$. Then by Theorem 2.14 and Heine-Borel Theorem, we see that we can extract from the sequence $f_n(z)(n = 1, 2, \dots)$ a subsequence $f_{n_k}(z)(k = 1, 2, \dots)$ which is uniformly convergent in the circle c'.

$2°$ The sequence $f_n(z_0)(n = 1, 2, \dots)$ is unbounded. Then we can find a subsequence $f_{n_k}(z_0)(k = 1, 2, \dots)$ such that

$$\lim_{k \to \infty} |f_{n_k}(z_0)| = \infty .$$

We are going to show that, as $k \to \infty$, $f_{n_k}(z)$ tends uniformly to infinity in the circle c'. In fact, consider a point z_1 of the circle c'. Then the circle $|z - z_1| < \frac{3}{4}R$ is interior to the circle c. By the same method of proof of (2.61), we get

$$\log |f_n(z) - a| < \frac{|b - a|R_1}{R_1 - |z - z_1|} \left(A \log^+ \frac{|f_n(z_1) - a|}{|b - a|} + B \log \frac{2R_1}{R_1 - |z - z_1|} \right)$$

where $R_1 = \frac{3}{4}R$. In particular in the circle $|z - z_1| < R/4$, we have

$$\log |f_n(z) - a| < \frac{3}{2}|b - a| \left(A \log^+ \frac{|f_n(z_1)| + |a|}{|b - a|} + B \log 3 \right) .$$

Since z_0 is a point of the circle $|z - z_1| < R/4$, we have

$$\log |f_{n_k}(z_0) - a| < \frac{3}{2}|b - a| \left(A \log^+ \frac{|f_{n_k}(z_1)| + |a|}{|b - a|} + B \log 3 \right) .$$

This inequality being true for any point z_1 of the circle c', it follows that $f_{n_k}(z)$ tends uniformly to infinity in the circle c' as $k \to \infty$.

Corollary 2.3. Let $\{f(z)\}$ be a family of holomorphic functions in a domain D. If each function $f(z)$ of the family $\{f(z)\}$ does not take two finite values $a(f)$ and $b(f)$ such that

$$|a(f)| < M , \quad |b(f)| < M , \quad |a(f) - b(f)| > \delta \tag{2.62}$$

where $M > 0, \delta > 0$ are two constants independent of $f(z)$, then the family $\{f(z)\}$ is normal in domain D.

Proof. Let $f_n(z)(n = 1, 2, \dots)$ be a sequence of the family $\{f(z)\}$. Set

$$a_n = a(f_n), \quad b_n = b(f_n) \quad (n = 1, 2, \dots)$$

and consider the sequence

$$g_n(z) = \frac{f_n(z) - a_n}{b_n - a_n} \quad (n = 1, 2, \dots) . \tag{2.63}$$

$g_n(z)$ does not take the values 0 and 1 in D, hence by Theorem 2.16, the family $\{g_n(z)(n = 1, 2, \dots)\}$ is normal in D, consequently we can get a subsequence $g_{n_k}(z)(k = 1, 2, \dots)$ satisfying one of the two conditions in Definition 2.3. If $g_{n_k}(z)(k = 1, 2, \dots)$ satisfies the first of these two conditions, then from (2.62) and (2.63), the sequence $f_{n_k}(z)(k = 1, 2, \dots)$ is locally uniformly bounded in D and then by Theorem 2.14, we can extract from the sequence $f_{n_k}(z)(k = 1, 2, \dots)$ a subsequence $f_{m_l}(z)(l = 1, 2, \dots)$ which is locally uniformly convergent in D. If $g_{n_k}(z)(k = 1, 2, \dots)$ satisfies the second condition in Definition 2.3, then evidently $f_{n_k}(z)(k = 1, 2, \dots)$ also satisfies the same condition.

We shall need also the following lemma:

Lemma 2.9. Let $\{f(z)\}$ be a normal family of holomorphic functions in a domain D. Let $f_n(z)(n = 1, 2, \ldots)$ be a sequence of this family. If for a point z_0 of D, the sequence $f_n(z_0)(n = 1, 2, \ldots)$ is bounded, then the sequence $f_n(z)(n = 1, 2, \ldots)$ is locally uniformly bounded in D.

Proof. Let ς_0 be a point of D, and let $c : |z - \varsigma_0| \leq r$ be a circle interior to D. It is sufficient to show that the sequence $f_n(z)(n = 1, 2, \ldots)$ is uniformly bounded in c. In fact, if this is untrue, then to each positive integer p corresponds a positive integer m_p such that

$$\max_{z \in c} |f_{m_p}(z)| > p .$$

From the sequence $g_p(z) = f_{m_p}(z)(p = 1, 2, \ldots)$ we can extract a subsequence $g_{p_s}(z)(s = 1, 2, \ldots)$ satisfying one of the two conditions in Definition 2.3. Since the sequence $g_{p_s}(z_0)(s = 1, 2, \ldots)$ is bounded, the sequence $g_{p_s}(z)(s = 1, 2, \ldots)$ must satisfy the first condition in Definition 2.3. But this is incompatible with the fact $\max_{z \in c} |g_{p_s}(z)| > p_s$.

2.6. FATOU'S THEORY ON THE FIX-POINTS OF ENTIRE FUNCTIONS

Let $f(z)$ be a transcendental entire function and let the sequence $f_n(z)$ $(n = 1, 2, \ldots)$ be defined as follows:

$$f_1(z) = f(z), \quad f_n(z) = f\{f_{n-1}(z)\} \quad (n = 2, 3, \ldots) .$$

Consider a point z_0 at which the family $\{f_n(z)(n = 1, 2, \ldots)\}$ is not normal, in other words, there does not exist a circle $|z - z_0| < r$ in which the family $\{f_n(z)(n = 1, 2, \ldots)\}$ is normal, then z_0 is called a Julia point of the family $\{f_n(z)(n = 1, 2, \ldots)\}$. The set of all Julia points of the family $\{f_n(z)(n = 1, 2, \ldots)\}$ is called the Julia set of the function $f(z)$ and is denoted by $J(f)$. The set $J(f)$ plays an important role in Fatou's theory. It will be shown that $J(f)$ is non-empty and has other properties. At present we first prove some simple properties of the set $J(f)$.

Theorem 2.17. If a point $z_0 \in J(f)$, then $f(z_0) \in J(f)$.

Proof. Assume that $f(z_0) \notin J(f)$. Then there is a circle $c : |z - f(z_0)| < R$ in which the family $\{f_n(z)(n = 1, 2, \ldots)\}$ is normal. Let $\gamma : |z - z_0| < r$ be a circle such that in γ we have

$$|f(z) - f(z_0)| < R/2 . \tag{2.64}$$

Consider a sequence $f_{n_k}(z)(k = 1, 2, \ldots; n_k \geq 2)$ of the family $\{f_n(z)(n = 1, 2, \ldots)\}$. By assumption, we can extract from the sequence $f_{n_k-1}(z)(k = 1, 2, \ldots)$ a subsequence $f_{m_h-1}(z)(h = 1, 2, \ldots)$ which converges uniformly in the circle $|z - f(z_0)| < R/2$ to a holomorphic function $g(z)$ or to the constant ∞. By (2.64), evidently in the circle γ, the sequence $f_{m_h}(z) = f_{m_h-1}\{f(z)\}(h = 1, 2, \ldots)$ converges uniformly to the holomorphic function $g\{f(z)\}$ or to the constant ∞. Hence the family $\{f_n(z)(n = 1, 2, \ldots)\}$ is normal in the circle γ, and we get a contradiction.

Theorem 2.18. If a point $z_0 \in J(f)$ and z_1 is a point such that $f(z_1) = z_0$ then $z_1 \in J(f)$.

Proof. Assume that $z_1 \notin J(f)$. Then there is a circle $c_1 : |z - z_1| < r_1$ such that the family $\{f_n(z)(n = 1, 2, \ldots)\}$ is normal in c_1. Let $\gamma : |z - z_0| < \rho$ be a circle such that the values taken by the function $f(z)$ in the circle $|z - z_1| < r_1/2$ cover γ. Now consider a sequence $f_{n_k}(z)(k = 1, 2, \ldots)$ of the family $\{f_n(z)(n = 1, 2, \ldots)\}$. By assumption, we can extract from the sequence $f_{n_k+1}(z)(k = 1, 2, \ldots)$ a subsequence $f_{m_h+1}(z)(h = 1, 2, \ldots)$ which converges uniformly in the circle $|z - z_1| < r_1/2$ to a holomorphic function or to the constant ∞. In the first case, given arbitrarily a positive number ε, we can find a positive integer H such that when $h \geq H, h' \geq H$, the inequality

$$|f_{m_h+1}(z) - f_{m_{h'}+1}(z)| < \varepsilon$$

holds in the circle $|z - z_1| < r_1/2$. Since

$$f_{m_h+1}(z) = f_{m_h}\{f(z)\}, \quad f_{m_{h'}+1}(z) = f_{m_{h'}}\{f(z)\}$$

and by the choice of the circle γ, we have

$$|f_{m_h}(z) - f_{m_{h'}}(z)| < \varepsilon$$

in γ. Hence in γ, the sequence $f_{m_h}(z)(h = 1, 2, \ldots)$ converges uniformly to a holomorphic function. In the second case, we see in the same way, that the sequence $f_{m_h}(z)(h = 1, 2, \ldots)$ converges uniformly to the constant ∞ in γ. Consequently the family $\{f_n(z)(n = 1, 2, \ldots)\}$ is normal in γ, and we get a contradiction.

Since the set $J(f)$ has the properties in Theorem 2.17 and 2.18, we say that $J(f)$ is completely invariant under the substitution $(z, f(z))$.

Theorem 2.19. $J(f_m) = J(f)$ $(m = 2, 3, \ldots)$.

Proof. Let $m \geq 2$ be an integer and set $g(z) = f_m(z)$. Then we have

$$g_1(z) = g(z) = f_m(z), \quad g_2(z) = g\{g_1(z)\} = f_m\{f_m(z)\} = f_{2m}(z) ,$$

and in general

$$g_n(z) = f_{nm}(z) \quad (n = 1, 2, \ldots) . \tag{2.65}$$

Let $z_0 \in J(g)$ and assume $z_0 \notin J(f)$. Then there is a circle $c : |z - z_0| < r$ in which the family $\{f_n(z)(n = 1, 2, \ldots)\}$ is normal. Since the functions of the family $\{g_n(z)(n = 1, 2, \ldots)\}$ belong to the family $\{f_n(z)(n = 1, 2, \ldots)\}$, hence the family $\{g_n(z)(n = 1, 2, \ldots)\}$ is also normal in c, and we get a contradiction.

Now let $z_0 \in J(f)$ and assume $z_0 \notin J(g)$. Then there is a circle $c :$ $|z - z_0| < r$ in which the family $\{g_n(z)(n = 1, 2, \ldots)\}$ is normal. By Corollary 2.2, the function $f_2(z)$ has an infinite number of fix points. Let α and β $(\alpha \neq \beta)$ be two fix-points of $f_2(z)$. Since the family $\{f_n(z)(n = 1, 2, \ldots)\}$ is not normal in c, by Theorem 2.16, there is a function $f_p(z)$ of this family, which takes at least one of the two values α ane β in c. Let this value be α, so that there is a point ς of c, such that

$$f_p(\varsigma) = \alpha .$$

Consider an integer $n \geq p$, then $n = p + 2h$ or $n = p + 2h + 1$ where $h \geq 0$ is an integer. In the first case

$$f_n(\varsigma) = f_{p+2h}(\varsigma) = f_{2h}\{f_p(\varsigma)\} = f_{2h}(\alpha) = \alpha$$

and in the second case,

$$f_n(\varsigma) = f_{p+2h+1}(\varsigma) = f_{2h+1}\{f_p(\varsigma)\} = f_{2h+1}(\alpha) = f(\alpha) .$$

Hence the sequence $f_n(\varsigma)(n = 1, 2, \ldots)$ is bounded and *a fortiori*, the sequence $g_n(\varsigma) = f_{nm}(\varsigma)(n = 1, 2, \ldots)$ is also bounded. By Lemma 2.9, the sequence $g_n(z)(n = 1, 2, \ldots)$ is uniformly bounded in the circle $c' : |z - z_0| \leq r/2$, accordingly in c' we have

$$|g_n(z)| \leq R \quad (n = 1, 2, \ldots) \tag{2.66}$$

where $R > 0$ is a constant independent of n and z.

Now let $n \geq m$, then $\lambda m \leq n < (\lambda + 1)m$ where $\lambda \geq 1$ is a positive integer, and we have

$$n = \lambda m + \mu, \quad 0 \leq \mu \leq m - 1 .$$

Next we have

$$f_n(z) = f_{\lambda m + \mu}(z) = f_\mu\{f_{\lambda m}(z)\} = f_\mu\{g_\lambda(z)\} . \tag{2.67}$$

From (2.66) and (2.67), we deduce that the sequence $f_n(z)(n = 1, 2, \ldots)$ is uniformly bounded in the circle c', this contradicts the condition $z_0 \in J(f)$.

Now we are going to make a classification of fix-points. In what follows, $f(z), f_n(z)$ and $J(f)$ have the same meaning as above.

If z_0 is a fix-point of the function $f(z)$, that is to say

$$f(z_0) = z_0 ,$$

then three cases are possible.

1) $|f'(z_0)| < 1$. In this case, the fix-point z_0 is said to be attractive.

2) $|f'(z_0)| > 1$. In this case, the fix-point z_0 is said to be repulsive.

3) $|f'(z_0)| = 1$. In this case, the fix-point z_0 is said to be neutral.

In the third case, we have $f'(z_0) = e^{2\pi\theta i}$, where $0 \leq \theta < 1$ is a real number, so we can again distinguish two cases:

a) θ is a rational number.

b) θ is an irrational number.

For the three kinds of fix-points, we have respectively the following theorems.

Theorem 2.20. If z_0 is an attractive fix-point of $f(z)$, then we can find a circle $c : |z - z_0| < r$ in which the sequence $f_n(z)(n = 1, 2, \ldots)$ converges uniformly to z_0.

Proof. By hypothesis, we have $|f'(z_0)| < 1$. Take a number h such that

$$|f'(z_0)| < h < 1$$

and let r be a positive number such that for $0 < |z - z_0| < r$, we have

$$\left| \frac{f(z) - z_0}{z - z_0} \right| = \left| \frac{f(z) - f(z_0)}{z - z_0} \right| < h .$$

Then in the circle $c : |z - z_0| < r$, we have

$$|f(z) - z_0| \leq h|z - z_0| \,,$$

and *a fortiori*,

$$|f(z) - z_0| \leq hr < r \,.$$

This shows that when $z \in c$, we have $f(z) \in c$. Hence in c, we have

$$|f_2(z) - z_0| \leq h|f(z) - z_0| \leq h^2|z - z_0| \,,$$
$$|f_3(z) - z_0| \leq h^3|z - z_0| \,,$$

and in general, in c we have

$$|f_n(z) - z_0| \leq h^n|z - z_0| \quad (n = 1, 2, \ldots) \,.$$

Hence in c the sequence $f_n(z)(n = 1, 2, \ldots)$ converges uniformly to z_0.

Theorem 2.21. If z_0 is a repulsive fix-point of $f(z)$, then $z_0 \in J(f)$.

Proof. By hypothesis, we have $|f'(z_0)| > 1$. We have

$$f_2'(z_0) = f'\{f(z_0)\}f'(z_0) = \{f'(z_0)\}^2 \,,$$
$$f_3'(z_0) = f'\{f_2(z_0)\}f_2'(z_0) = \{f'(z_0)\}^3 \,,$$

and in general

$$f_n'(z_0) = \{f'(z_0)\}^n \quad (n = 1, 2, \ldots) \,. \tag{2.68}$$

Assume, on the contrary, there is a circle $c : |z - z_0| < r$, in which the family $\{f_n(z)(n = 1, 2, \ldots)\}$ is normal. Since

$$f_n(z_0) = z_0 \quad (n = 1, 2, \ldots) \,, \tag{2.69}$$

by Lemma 2.8, we can get a positive number M such that in the circle $|z - z_0| < r/2$, we have

$$|f_n(z)| \leq M \quad (n = 1, 2, \ldots) \,.$$

Then by Cauchy's inequality, we have

$$\frac{r}{2}|f_n'(z_0)| = \frac{r}{2}|f'(z_0)|^n \leq M \,,$$

which is impossible, because $|f'(z_0)| > 1$.

Lemma 2.10. If z_0 is a fix-point of $f(z)$ such that $f'(z_0) = 1$, then $z_0 \in J(f)$.

Proof. By hypothesis, we have

$$f(z) = z_0 + (z - z_0) + a(z - z_0)^m + b(z - z_0)^{m+1} + \ldots \qquad (2.70)$$

where $m \geq 2, a \neq 0$. We are going to show that in general we have

$$f_n(z) = z_0 + (z - z_0) + na(z - z_0)^m + b_n(z - z_0)^{m+1} + \ldots . \qquad (2.71)$$

In fact, if (2.71) holds for an integer n, then

$$f_n(z) - z_0 = (z - z_0)\{1 + na(z - z_0)^{m-1}\psi(z)\} , \quad \psi(z_0) = 1 ,$$

and hence

$$f_{n+1}(z) - z_0 = f_n\{f(z)\} - z_0 = \{f(z) - z_0\}\{1 + na[f(z) - z_0]^{m-1}\Psi(z)\} ,$$
$$\Psi(z_0) = 1 .$$

On the other hand, by (2.70), we have

$$f(z) - z_0 = (z - z_0)\{1 + a(z - z_0)^{m-1}\varphi(z)\}, \quad \varphi(z_0) = 1 ,$$

hence

$$f_{n+1}(z) - z_0 = (z - z_0)\{1 + a(z - z_0)^{m-1}\varphi(z)\}$$
$$\times \{1 + na(z - z_0)^{m-1}[1 + a(z - z_0)^{m-1}\varphi(z)]^{m-1}\Psi(z)\}$$
$$= (z - z_0)\{1 + (z - z_0)^{m-1}\varphi(z)\}\{1 + na(z - z_0)^{m-1}\varphi_1(z)\} ,$$
$$\varphi_1(z_0) = 1 .$$

Hence

$$f_{n+1}(z) - z_0 = (z - z_0)\{1 + (z - z_0)^{m-1}\Phi(z)\} , \quad \Phi(z_0) = (n+1)a .$$

So (2.71) also holds for $n + 1$. Finally by the method used for the proof of Theorem 2.21, we see that $z_0 \in J(f)$.

Theorem 2.22. If z_0 is a fix-point of $f(z)$ such that $f'(z_0) = e^{2\pi\theta i}$, where $\theta(0 \leq \theta < 1)$ is a rational number, then $z_0 \in J(f)$.

Proof. By Lemma 2.10, we may assume $0 < \theta < 1, \theta = p/q$, where p and q are two positive integers. By (2.68), we have

$$f'_q(z_0) = \{f'(z_0)\}^q = e^{2\pi p i} = 1 ,$$

and by (2.69), z_0 is a fix-point of $f_q(z)$, hence by Lemma 2.10, $z_0 \in J(f_q)$ and then by Theorem 2.19, $z_0 \in J(f)$.

It remains to study neutral fix-points for the case b). For this purpose, we need the following lemma:

Lemma 2.11. Let a and b be two real numbers and let $\varepsilon > 0$ be a number. Then there exist two integers, m and n not both equal to zero, such that

$$|ma + nb| < \varepsilon . \tag{2.72}$$

Proof. Assume, on the contrary, that such two integers m and n do not exist. Set $X_{m,n} = ma + nb$. Then we have

$$|X_{m,n}| \geq \varepsilon$$

for any pair of integers (positive, negative or zero) (m, n) not both equal to zero. For two such pairs of integers (m_1, n_1) and (m_2, n_2) distinct from each other, we have also

$$|X_{m_1,n_1} - X_{m_2,n_2}| = |(m_1 - m_2)a + (n_1 - n_2)b| \geq \varepsilon . \tag{2.73}$$

Now take an integer $N > 1$ and consider the set S of the pairs of integers (p, q) such that

$$1 \leq p \leq N, \quad 1 \leq q \leq N .$$

The total number of such pairs (p, q) is N^2. To each pair (p, q) of S, corresponds the interval

$$I_{p,q} : \left[X_{p,q} - \frac{\varepsilon}{3}, X_{p,q} + \frac{\varepsilon}{3} \right] .$$

Evidently

$$I_{p,q} \subset \left[- \left(\frac{\varepsilon}{3} + 2NA \right), \ \frac{\varepsilon}{3} + 2NA \right]$$

where $A = \max(|a|, |b|)$. On the other hand, by (2.73), corresponding to two distinct pairs (p_1, q_1) and (p_2, q_2) of S, the intervals $I_{p_1 q_1}$ and I_{p_2,q_2} have no common point. Consequently the sum of the lengths of the N^2 intervals $I_{p,q}$ is less than the length of the interval $\left[- \left(\frac{\varepsilon}{3} + 2NA \right), \frac{2}{3} + 2NA \right]$, hence

$$\frac{2}{3} N^2 \varepsilon < 2 \left(\frac{\varepsilon}{3} + 2NA \right) ,$$

$$\frac{2}{3} \varepsilon < \frac{4NA}{N^2 - 1} .$$

Since the right member of this inequality tends to zero, as $N \to \infty$, we get a contradiction.

Theorem 2.23. Let z_0 be a fix-point of $f(z)$. If $f'(z_0) = e^{2\pi\theta i}$ where θ is an irrational number such that $0 < \theta < 1$, and if $z_0 \notin J(f)$, then there exist a circle $c : |z - z_0| < r$ and a sequence of positive integers $\lambda_p (p = 1, 2, \ldots)$ such that in c, the sequence $f_{\lambda_p}(z)(p = 1, 2, \ldots)$ converges uniformly to z.

Proof. Apply Lemma 2.11 to the case $a = \theta, b = 1$ then for each positive integer k, there are two integers m_k and n_k not both equal to zero, such that

$$|m_k\theta + n_k| < 1/k . \tag{2.74}$$

In this inequality, m_k cannot be equal to zero, because if $m_k = 0$, then $n_k \neq 0$ and $|n_k| < 1/k$. We may assume $m_k > 0$. Consequently there exists a sequence of positive integers $m_k (k = 1, 2, \ldots)$ such that

$$\lim_{k \to \infty} e^{2\pi m_k \theta i} = 1 . \tag{2.75}$$

On the other hand, by hypothesis, there is a circle $c : |z - z_0| < R$ in which the family $\{f_n(z)(n = 1, 2, \ldots)\}$ is normal. Since $f_n(z_0) = z_0(n = 1, 2, \ldots)$, we can extract from the sequence $f_{m_k}(z)(k = 1, 2, \ldots)$ a subsequence $f_{\lambda_p}(z)(p = 1, 2, \ldots)$ which converges uniformly to a holomorphic function $g(z)$ in the circle $\gamma : |z - z_0| < r$ $(r = R/2)$. We are going to show that $g(z) = z$.

First of all, by (2.75) and the formula

$$f'_{\lambda_p}(z_0) = \{f'(z_0)\}^{\lambda_p} ,$$

we have

$$g'(z_0) = \lim_{p \to \infty} f'_{\lambda_p}(z_0) = 1 .$$

Now assume, on the contrary, that the identity $g(z) = z$ is not satisfied. Then in the circle γ, we have

$$g(z) = z_0 + (z - z_0) + b(z - z_0)^m + b'(z - z_0)^{m+1} + \ldots \tag{2.76}$$

where $b \neq 0, m \geq 2$. On the other hand, we have

$$f(z) = z_0 + \mu(z - z_0) + a_2(z - z_0)^2 + \ldots + a_n(z - z_0)^n + \ldots \tag{2.77}$$

where

$$\mu = e^{2\pi\theta i} .$$

In the formula

$$f\{f_{\lambda_p}(z)\} = f_{\lambda_p}\{f(z)\} ,$$

let $p \to \infty$, we get

$$f\{g(z)\} = g\{f(z)\} .$$

Both sides of this formula are holomorphic functions in a circle $\gamma' : |z-z_0| < r'$ $(0 < r' < r)$. Let us find out the coefficients of the term $(z - z_0)^m$ in the expansions of $f\{g(z)\}$ and $g\{f(z)\}$.

First by (2.76), we have

$$g(z) - z_0 = (z - z_0)\{1 + b(z - z_0)^{m-1}\varphi(z)\}, \quad \varphi(z_0) = 1 . \tag{2.78}$$

Next by (2.77), we have

$$f\{g(z)\} = z_0 + \mu\{g(z) - z_0\} + \ldots + a_n\{g(z) - z_0\}^n + \ldots . \tag{2.79}$$

Noting that

$$\begin{aligned}
\{g(z) - z_0\}^n &= (z - z_0)^n\{1 + b(z - z_0)^{m-1}\varphi(z)\}^n \\
&= (z - z_0)^n\{1 + (z - z_0)^{m-1}\varphi_n(z)\} \\
&= (z - z_0)^n + (z - z_0)^{n+m-1}\varphi_n(z) \tag{2.80}
\end{aligned}$$

and for $n \geq 2$, we have $n + m - 1 \geq m + 1$, we see from (2.79) and (2.80), that in the expansion of $f\{g(z)\}$ the coefficient of the term $(z - z_0)^m$ is $a_m + \mu b$. On the other hand, by (2.77), we have

$$f(z) - z_0 = \mu(z - z_0)\psi(z), \quad \psi(z_0) = 1 .$$

Then by (2.78), we have

$$g\{f(z)\} - z_0 = \{f(z) - z_0\} + b\mu^m(z - z_0)^m\Psi(z), \quad \Psi(z_0) = 1 .$$

Hence in the expansion of $g\{f(z)\}$ the coefficient of the term $(z - z_0)^m$ is $a_m + b\mu^m$. Consequently

$$a_m + \mu b = a_m + b\mu^m ,$$
$$\mu^{m-1} = 1 .$$

This is impossible, because θ is an irrational number. So we arrive at a contradiction.

Concerning the set $J(f)$ the following theorem of Fatou is fundamental which shows the relationship between $J(f)$ and the fix-points.

Theorem 2.24. The set $J(f)$ has the following properties:

$1°$ $J(f)$ consists of an infinite number of points and is unbounded.

$2°$ $J(f)$ is a perfect set, in other words, $J(f) = \{J(f)\}'$.

$3°$ Let E_n be the set of the fix-points of $f_n(z)$ and define

$$E(f) = \bigcup_{n=1}^{\infty} E_n .$$

Then each point of $J(f)$ is a point of accumulation of $E(f)$.

For the sake of convenience, let us recall the definition of the notion of point of accumulation. Let S be a set of points of the complex plane. A point z_0 is called a point of accumulation of S, if in each circle $|z - z_0| < r$ there is a point $z' \neq z_0, z' \in S$. The set of the points of accumulation of S is denoted by S'. If $S = S'$, then S is said to be perfect (S is assumed non-empty).

For the proof of Theorem 2.24, we need the following lemma:

Lemma 2.12. Let D be a domain. Assume that there exist three distinct points a_1, a_2 and b of D, such that

1) a_1 and a_2 are fix-points of $f(z)$.

2) $f(b)$ is equal to one of a_1 and a_2.

Then there is a point z_0 of D such that $z_0 \in J(f)$.

Proof. Distinguish four cases:

$1°$ Among the two points a_1 and a_2, there is at least one, say a_1, which is a repulsive fix point of $f(z)$. Then by Theorem 2.21, $a_1 \in J(f)$.

$2°$ Among the two points a_1 and a_2, there is at least one, say a_1, which is a neutral fix-point of $f(z)$ and belongs to the kind a). Then by Theorem 2.2, $a_1 \in J(f)$.

$3°$ a_1 is an attractive fix-point of $f(z)$. Then by Theorem 2.20, there is a circle $c : |z - a_1| < r$ in which the sequence $f_n(z)(n = 1, 2, \ldots)$ converges uniformly to a_1. We may suppose that $c \subset D$. Now assume, on the contrary, that there does not exist a point z_0 of D such that $z_0 \in J(f)$. Then by Theorem 2.15, the family $\{f_n(z)(n = 1, 2, \ldots)\}$ is normal in D. Consequently we can get a subsequence $f_{n_k}(z)(k = 1, 2, \ldots)$ which converges locally uniformly to a holomorphic function $g(z)$. Since in the circle c, we have $g(z) = a_1$, hence $g(z) = a_1$ in D. On the other hand, from $f_n(a_2) = a_2$ $(n = 1, 2, \ldots)$, we have $g(a_2) = a_2$. So we get contradiction.

$4°$ a_1 is a neutral fix-point of $f(z)$ and belongs to the kind b). Assume, on the contrary, that there does not exist a point z_0 of D such that $z_0 \in J(f)$. Then the family $\{f_n(z)(n = 1, \ldots 2, \ldots)\}$ is normal in D. By Theorem 2.23, we can get a circle $c : |z - a_1| < r$ interior to D and a sequence of positive integers $\lambda_p(p = 1, 2, \ldots)$, such that the sequence $f_{\lambda_p}(z)(p = 1, 2, \ldots)$ converges uniformly to z in c. From the sequence $f_{\lambda_p}(z)(p = 1, 2, \ldots)$ we can extract a subsequence $f_{\mu_q}(z)(q = 1, 2, \ldots)$ which converges locally uniformly to a holomorphic function $g(z)$ in D. Since $g(z) = z$ in the circle c, hence $g(z) = z$ in D. In particular, $g(b) = b$. On the other hand, by hypothesis, $f(b)$ is equal to one of a_1 and a_2, say $f(b) = a_2$, hence

$$f_2(b) = f\{f(b)\} = f(a_2) = a_2 ,$$
$$f_3(b) = f\{f_2(b)\} = f(a_2) = a_2 ,$$

and in general,

$$f_n(b) = a_2 \quad (n = 1, 2, \ldots) . \tag{2.81}$$

Hence $g(b) = a_2$ and we get a contradiction.

Now let us return to the proof of Theorem 2.24. First suppose that $f(z)$ has an infinite number of fix-points. To prove the part $1°$ of Theorem 2.24, consider a positive number R. Since the function $f(z) - z$ has an infinite number of zeros, let a_1 and a_2 be its two distinct zeros such that $|a_j| > R$ $(j = 1, 2,)$. Again since $f(z)$ is a transcendental entire function, one at least of a_1 and a_2, say a_1, is such that $f(z) - a_1$ has an infinite number of zeros, hence $f(z) - a_1$ has a zero b with $|b| > R$, $b \neq a_j$ $(j = 1, 2)$. Take a number $R' > \max(|a_1|, |a_2|, |b|)$, then in the domain $\Gamma : R < |z| < R'$ there are three distinct points a_1, a_2, b satisfying the conditions 1) and 2) in Lemma 2.12. Consequently in Γ there is a point $z_0 \in J(f)$, and the part $1°$ in Theorem 2.24 is proved.

To prove the part $2°$ in Theorem 2.24, we first show that $J(f) \subset \{J(f)\}'$. Consider a point $z_0 \in J(f)$ and a circle $c : |z - z_0| < r$. We are going to show that in c there is a point z' such that

$$z' \neq z_0, \quad z' \in J(f) . \tag{2.82}$$

In fact, by hypothesis the family $\{f_n(z)(n = 1, 2, \ldots)\}$ is not normal in c. Consequently there is a subsequence $f_{n_k}(z)(k = 1, 2, \ldots)$ satisfying the following property: From the sequence $f_{n_k}(z)(k = 1, 2, \ldots)$ we cannot extract a subsequence which converges locally uniformly in c to a holomorphic function or to the constant ∞. Consider the sequence $f_{n_k}(z_0)(k = 1, 2, \ldots)$ from which we can extract a subsequence $f_{m_h}(z_0)(h = 1, 2, \ldots)$ converging to a limit λ finite or infinite. Evidently the sequence $f_{m_h}(z)$ $(h = 1, 2, \ldots)$ has also the above property. From the sequence $f_{m_h}(z)(h = 1, 2, \ldots)$ we cannot extract a subsequence which converges locally uniformly in c to a holomorphic function or to the constant ∞. Now distinguish two cases:

1) λ is finite. Take two domains $\Gamma : R < |z| < R'$ and $\Gamma_1 : R_1 < |z| < R_1'$ such that $|\lambda| < R < R_1$ and that in Γ there is a point $\varsigma_0 \in J(f)$ and in Γ_1 a point $\varsigma_1 \in J(f)$. This is possible by part 1° in Theorem 2.24. Let H be a positive integer such that

$$|f_{m_h}(z_0)| < R \quad \text{for } h \geq H . \tag{2.83}$$

By Corollary 2.3 and the above property of the sequence $f_{m_h}(z)$ $(h = 1, 2, \ldots)$, there exists an integer $h \geq H$ such that the values taken by the function $f_{m_h}(z)$ in c cover one of the domains Γ or Γ_1, say Γ. Consequently there is a point z' of c such that $f_{m_h}(z') = \varsigma_0$. By Theorem 2.19, $\varsigma_0 \in J(f_{m_h})$, and then by Theorem 2.18, $z' \in J(f_{m_h}) = J(f)$. Hence z' satisfies the condition (2.82).

2) λ is infinite. Again take two domains $\Gamma : R < |z| < R'$ and $\Gamma_1 : R_1 < |z| < R_1'$ such that $R' < R_1$ and that in Γ there is a point $\varsigma_0 \in J(f)$ and in Γ_1 a point $\varsigma_1 \in J(f)$. Next take a positive integer H such that

$$|f_{m_h}(z_0)| > R_1' \quad \text{for } h \geq H \tag{2.84}$$

and then in the same way as in the above case, we see that there is a point z' of c satisfying the condition (2.82).

In the above we have shown that $J(f) \subset \{J(f)\}'$. Conversely it is evident that $\{J(f)\}' \subset J(f)$. The part 2° in Theorem 2.24 is proved.

It remains to prove the part 3° of Theorem 2.24. Let $z_0 \in J(f)$. We know already that z_0 is a point of accumulation of $J(f)$. Consider a circle $c : |z - z_0| < r$. Then we can get four distinct points α_j $(j = 1, 2, 3, 4)$ of c, such that

$$\alpha_j \neq z_0, \quad \alpha_j \in J(f) \; (j = 1, 2, 3, 4) .$$

By the complement of Corollary 1.6, among the four values α_j ($j = 1, 2, 3, 4$) there are at most two which are completely multiple values of $f(z)$. So we may assume that α_1, α_2 are not completely multiple values of $f(z)$. Among the two values α_1, α_2, there is at least one, say α_1, such that the function $f(z) - \alpha_1$ has an infinite number of zeros. There is one such zero β whose order is 1, namely,

$$f(\beta) = \alpha_1, \quad f(\beta) \neq 0 .$$

Consequently we can get a circle $\gamma : |z - \alpha_1| < \rho$ and a holomorphic function $g(z)$ in γ, satisfying

$$f\{g(z)\} = z \tag{2.85}$$

in γ. We may assume that ρ is sufficiently small, such that γ is interior to c and that γ does not contain the point z_0. We are going to show that there exists a point z_1 of γ such that

$$z_1 \in E(f) . \tag{2.86}$$

Distinguish two cases:

1) There is a point z_1 of γ such that $g(z_1) = z_1$. Then $f(z_1) = z_1$, and hence z_1 satisfies (2.86).

2) In γ, there is no point z_1 such that $g(z_1) = z_1$. Then the functions

$$F_n(z) = \frac{f_n(z) - z}{g(z) - z} \quad (n = 1, 2, \ldots)$$

are holomorphic in γ. Assume, on the contrary, that in γ there is no point z_1 satisfying (2.86). Then by (2.85), we see that in γ the functions $F_n(z)(n = 1, 2, \ldots)$ do not take the values 0 and 1. It follows that the family $\{F_n(z)(n = 1, 2, \ldots)\}$ is normal in γ. This implies that the family $\{f_n(z)(n = 1, 2, \ldots)\}$ is normal in γ. But $\alpha_1 \in J(f)$ so we get a contradiction.

Thus we have shown that there exists a point z_1 of c, such that

$$z_1 \neq z_0 , \quad z_1 \in E(f) .$$

Hence z_0 is a point of accumulation of $E(f)$.

In the proof of Theorem 2.24 given above, we have assumed that the function $f(z)$ has an infinite number of fix-points. If this condition is not satisfied, we first apply Theorem 2.24 to the function $f_2(z)$. Since $J(f) = J(f_2)$, hence Theorem 2.24 still holds.

Theorem 2.24 shows that the set $J(f)$ is closely related to the fix-points of all orders of the function $f(z)$. So there is interest to study the distribution of $J(f)$ in the complex plane. In this respect we have the following interesting theorem.

Theorem 2.25. If the set $J(f)$ has an interior point, in other words, there exists a circle $c : |z - z_0| < r$ belonging to $J(f)$, then $J(f)$ is the complex plane.

For the proof of this theorem we need the following lemma:

Lemma 2.13. Let $z_0 \in J(f)$. Then for each finite value a, there exist a sequence of positive integers n_k $(k = 1, 2, \ldots)$ and a sequence of points $\varsigma_k (k = 1, 2, \ldots)$ such that

$$\lim_{k \to \infty} n_k = \infty, \quad \lim_{k \to \infty} \varsigma_k = z_0, \quad f_{n_k}(\varsigma_k) = a \ (k = 1, 2, \ldots), \tag{2.87}$$

except at most for one finite value a.

Proof. Consider a finite value a. If for each positive integer N, there exist a positive integer n and a point ς such that

$$n \geq N, \quad |\varsigma - z_0| < \frac{1}{N}, \quad f_n(\varsigma) = a \ ,$$

then this value a has the required property in Lemma 2.13. In fact, for each positive integer k, there exist a positive integer n_k and a point ς_k such that

$$n_k \geq k, \quad |\varsigma_k - z_0| < \frac{1}{k}, \quad f_{n_k}(\varsigma_k) = a \ .$$

Therefore if a finite value a does not have the required property in Lemma 2.13, then there exists a positive integer N such that for $n \geq N$, the function $f_n(z)$ does not take the value a in the circle $|z - z_0| < 1/N$. It follows that if there are two finite values a and a' $(a \neq a')$ both not having the required property in Lemma 2.13, then we can find a positive integer N_1 such that for $n \geq N_1$, the function $f_n(z)$ does not take the values a and a' in the circle $|z - z_0| < 1/N_1$. By Theorem 2.16, the family $\{f_n(z)(n \geq N_1)\}$ is normal in the circle $|z - z_0| < 1/N_1$. It is easy to see that the family $\{f_n(z)(n = 1, 2, \ldots)\}$ is then also normal in the circle $|z - z_0| < 1/N_1$. This contradicts the hypothesis $z_0 \in J(f)$.

Now let us prove Theorem 2.25. Assume that there is a circle $c : |z - z_0| < r$ belonging to $J(f)$. Let a be a finite value satisfying the condition (2.87) in Lemma 2.13. Then we can get a positive integer k such that $\varsigma_k \in c$. Hence $\varsigma_k \in J(f) = J(f_{n_k})$. Since $f_{n_k}(\varsigma_k) = a$, we have $a \in J(f_{n_k})$ by Theorem 2.17, and therefore $a \in J(f)$. It is then clear that every finite value $a \in J(f)$, and so $J(f)$ is the complex plane.

In view of Theorem 2.25, it is natural to ask whether we can find a transcendental entire function $f(z)$ such that $J(f)$ is the complex plane. The answer to this question is affirmative. For reference, see

1. I.N. Baker, "Limit functions and sets of non-normality in iteration theory", *Ann. Acad. Sci. Fenn. Ser. A.I. Math.* **467** (1970).

2. M. Misiurewicz, "On iterates of e^z", *Ergodic Theory Dynamical Systems* **1** (1981) 103-106.

If the set $J(f)$ is not the complex plane, then its complementary set $N(f)$ is non-empty. The set $N(f)$ is the set of points z_0 at which the family $\{f_n(z)(n = 1, 2, \ldots)\}$ is normal. Evidently $N(f)$ is an unbounded open set. From Theorems 2.17 and 2.18, it is easy to see that the set $N(f)$ is also completely invariant under the substitution $(z, f(z))$. A domain $D \subset N(f)$ is called a component of $N(f)$, if there does not exist a domain Δ such that

$$D \subset \Delta \subset N(f), \quad D \neq \Delta .$$

For reference concerning the set $J(f)$ and the components of the set $N(f)$, see

1. H. Töpfer, "Über die Iteration der ganzen transzendenten Funktionen, insbesondere von sin z und cos z", *Math. Ann.* **117** (1941) 65-84.

2. I.N. Baker, "Sets of non-normality in iteration theory", *J. London Math. Soc.* **40** (1965) 499-502.

3. I.N. Baker, "Repulsive fix-points of entire functions", *Math. Zeit.* **104** (1968) 252-256.

4. I.N. Baker, "The domains of normality of an entire function", *Ann. Acad. Sci. Fenn. Ser. AI* **1** (1975) 277-283.

5. L.S.O. Liverpool, "Value distribution and related questions in iteration theory", *Factorization Theory of Meromorphic Functions*, New York, 1982, pp. 55-69.

6. I.N. Baker, "Wandering domains in the iteration of entire functions", *Proc. London Math. Soc.* **49** (1984) 563-576.

7. A.K. Kromonko and M. Yu. Ljubič, "Iterates of entire functions",

Preprint of the Physico-Technical Institute of Low Temperature, UkrSSR Academy of Sciences, Kharkov, 1984, N6, pp. 1-37.

8. Chi-tai Chuang, "A simple proof of a theorem of Fatou on the iteration and fix-points of transcendental entire functions", *Contemporary Mathematics* **48** (1985).

9. I.N. Baker, "Iteration of entire functions: An introductory survey", *Lectures on Complex Analysis*, Singapore, 1988.

2.7. CASE OF POLYNOMIALS

In what precedes, we have obtained some results on the fix-points of a transcendental entire function and of its iterates. Most of them are also valid for the case of polynomials.

Consider a polynomial $P(z)$ of degree $d \geq 2$:

$$P(z) = a_d z^d + a_{d-1} z^{d-1} + \ldots + a_1 z + a_0 \quad (a_d \neq 0) \ .$$

Let $P_n(z)(n = 1, 2, \ldots)$ be the sequence defined as follows:

$$P_1(z) = P(z), \quad P_n(z) = P\{P_{n-1}(z)\} \quad (n = 2, 3, \ldots)$$

where $P_n(z)$ is a polynomial of degree d^n. Define $J(P)$ to be the set of points z_0 at which the family $\{P_n(z)(n = 1, 2, \ldots)\}$ is not normal. By the same method used in Sec. 2.6, we can prove the following two theorems:

Theorem 2.26. If a point $z_0 \in J(P)$, then $P(z_0) \in J(P)$.

Theorem 2.27. If a point $z_0 \in J(P)$ and z_1 is a point such that $P(z_1) = z_0$, then $z_1 \in J(P)$.

So the set $J(P)$ is completely invariant under the substitution $\{z, P(z)\}$.

Theorem 2.28. $J(P_m) = J(P)(m = 2, 3, \ldots)$.

Proof. Let $m \geq 2$ be an integer. By the same method used in the proof of Theorem 2.19, we see that if $z_0 \in J(P_m)$, then $z_0 \in J(P)$. Conversely let $z_0 \in J(P)$. To show that $z_0 \in J(P_m)$, we again use the method employed in the proof of Theorem 2.19. We see that it is sufficient to prove that at least one of the polynomials $P(z)$ or $P_2(z)$ has two fix-points α and β $(\alpha \neq \beta)$. To see this, assume, on the contrary, that each of $P(z)$ and $P_2(z)$ has only one fix-point (finite). Then

$$P(z) - z = A_1(z - z_1)^d, \quad P_2(z) - z = A_2(z - z_2)^{d^2}$$

where $A_j \neq 0$ $(j = 1, 2)$ are constants. It follows that

$$
\begin{aligned}
P_2(z) &= A_1\{P(z) - z_1\}^d + P(z) \\
&= A_1\{A_1(z - z_1)^d + z - z_1\}^d + A_1(z - z_1)^d + z , \\
P_2(z) - z &= A_1(z - z_1)^d\{[A_1(z - z_1)^{d-1} + 1]^d + 1\} .
\end{aligned} \tag{2.88}
$$

On the other hand, we have

$$
P_2(z) - z = A_2\{(z - z_1) + (z_1 - z_2)\}^{d^2} . \tag{2.89}
$$

The right members of (2.88) and (2.89) are polynomials of $z - z_1$. Since these two polynomials do not have the same coefficients, we get a contradiction.

Since the polynomial $P(z) - z$ has zeros, the polynomial $P(z)$ always has fix-points. As in the case of transcendental entire functions, a fix-point z_0 of $P(z)$ is said to be attractive, repulsive or neutral, according to $|P'(z_0)|$ is less than 1, greater than 1 or equal to 1. In the third case, we have $P'(z_0) = e^{2\pi\theta i}$, $(0 \leq \theta < 1)$; we distinguish again two cases according to the number θ is rational or irrational.

Moreover for these three kinds of fix-points of $P(z)$, we have correspondingly four theorems which are obtained from Theorems 2.20, 2.21, 2.22 and 2.23, in replacing $f(z)$ by $P(z)$.

Now we are going to prove the following theorem:

Theorem 2.29. There exists a number $R > 1$ such that in the domain $|z| > R$, the sequence $|P_n(z)|(n = 1, 2, \ldots)$ converges uniformly to ∞.

Proof. We have

$$
P(z) = a_d z^d\{1 + \varphi(z)\} .
$$

Since $d \geq 2$, we can find a number $R > 1$ such that for $|z| > R$, we have

$$
|a_d z^{d-1}\{1 + \varphi(z)\}| > 2 .
$$

Then for $|z| > R$, we have

$$
\begin{aligned}
|P(z)| &> 2|z| > 2R > R , \\
|P_2(z)| &> 2|P(z)| > 2^2|z| > 2^2 R > R , \\
|P_3(z)| &> 2|P_2(z)| > 2^3|z| > 2^3 R > R ,
\end{aligned}
$$

in general

$$
P_n(z) > 2^n R \quad (n = 1, 2, \ldots) .
$$

Hence in the domain $|z| > R$, the sequence $|P_n(z)|$ $(n = 1, 2, \dots)$ converges uniformly to ∞.

Theorem 2.29 shows that the point $z = \infty$ is one at which the family $\{P_n(z)(n = 1, 2, \dots)\}$ is normal. On the other hand, the point $z = \infty$ is a fix-point of $P(z)$ and it is attractive in the sense that the point $w = 0$ is an attractive fix-point of the function

$$\frac{1}{P\left(\frac{1}{w}\right)} = \frac{w^d}{a_0 w^d + a_1 w^{d-1} + \dots + a_{d-1} w + a_d} .$$

Finally we are going to prove the following theorem:

Theorem 2.30. The set $J(P)$ has the following properties:
1° $J(P)$ is a non-empty bounded perfect set.
2° Let E_n be the set of the fix-points of $P_n(z)$ and define

$$E(P) = \bigcup_{n=1}^{\infty} E_n .$$

Then each point of $J(P)$ is a point of accumulation of $E(P)$.

Proof. First we prove that $J(P)$ is non-empty. In fact, if $J(P)$ is empty, the family $\{P_n(z)(n = 1, 2, \dots)\}$ is normal in the complex plane. Let z_0 be a fix-point of $P(z)$. We have

$$P_n(z_0) = z_0 \quad (n = 1, 2, \dots) .$$

So the sequence $P_n(z_0)(n = 1, 2, \dots)$ is bounded. Then by Lemma 2.9, the sequence $P_n(z)(n = 1, 2, \dots)$ is locally uniformly bounded in the complex plane. This contradicts Theorem 2.29.

By Theorem 2.29, the set $J(P)$ belongs to the circle $|z| \leq R$ and therefore is bounded.

To prove that $J(P)$ is perfect, it is sufficient to show that $J(P) \subset \{J(P)\}'$. First of all, by Theorem 2.13, there exist three positive integers $n_i (i = 1, 2, 3)$ such that $n_1 < n_2 < n_3$ and that there are three points $\varsigma_i (i = 1, 2, 3)$ which are fix points of $P(z)$ of exact order $n_i (i = 1, 2, 3)$ respectively. Obviously $\varsigma_i (i = 1, 2, 3)$ are distinct. Since $P_{n_i}(\varsigma_i) = \varsigma_i (i = 1, 2, 3)$, we see that the sequences

$$P_n(\varsigma_i) \ (n = 1, 2, \dots) \quad (i = 1, 2, 3) \tag{2.90}$$

are bounded.

Now consider a point $z_0 \in J(P)$ and a circle $c : |z - z_0| < r$. We are going to show that in c there is a point z' such that

$$z' \neq z_0, \quad z' \in J(P) . \tag{2.91}$$

This is proved by the same method used in the proof of the part $2°$ in Theorem 2.24, with suitable modifications. Since the family $\{P_n(z)(n = 1, 2, \dots)\}$ is not normal in c, there is a subsequence $P_{n_k}(z)(k = 1, 2, \dots)$ such that from the sequence $P_{n_k}(z)(k = 1, 2, \dots)$ we cannot extract a subsequence which converges locally uniformly in c to a holomorphic function or to the constant ∞. Consider the sequence $P_{n_k}(z_0)(k = 1, 2, \dots)$. From this sequence we can extract a subsequence $P_{m_h}(z_0)(h = 1, 2, \dots)$ converging to a limit λ, finite or infinite. Distinguish two cases:

1) λ is finite. Among the three points ς_i $(i = 1, 2, 3)$ there are at least two points, say ς_1 and ς_2, not equal to λ. Choose two bounded domains $\Omega_j (j = 1, 2)$ satisfying the following conditions:

a) $\varsigma_i \in \Omega_i (i = 1, 2), \quad \Omega_i \cap D \neq \phi \ (i = 1, 2)$
where D is the domain $|z| > R$ in Theorem 2.29.

b) $\overline{\Omega}_1 \cap \overline{\Omega}_2 = \phi, \quad \lambda \notin \overline{\Omega}_i \ (i = 1, 2)$
where $\overline{\Omega}_i$ denotes the closure of Ω_i.

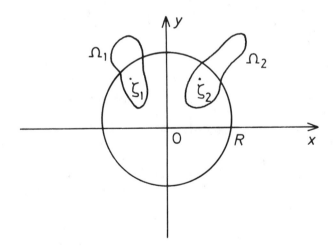

Then evidently we can get two positive numbers M and δ such that for any two points $z_i \in \Omega_i (i = 1, 2)$ we have

$$|z_i| < M \ (i = 1, 2,), \quad |z_1 - z_2| \geq \delta \tag{2.92}$$

and that

$$\overline{\Omega}_i \cap \gamma = \phi \quad (i = 1, 2) \tag{2.93}$$

where γ denotes the circle $|z - \lambda| < \delta$. In view of the condition a) and the boundedness of the sequences (2.90), we see that the family $\{P_n(z)(n = 1, 2, \ldots)\}$ is not normal in Ω_i. Hence there are two points $\alpha_i(i = 1, 2)$ such that

$$\alpha_i \in \Omega_i \ (i = 1, 2), \quad \alpha_i \in J(P) \ (i = 1, 2) . \tag{2.94}$$

Let H be a positive integer such that

$$P_{m_h}(z_0) \in \gamma \quad \text{for} \quad h \geq H . \tag{2.95}$$

As in the proof of the part 2° in Theorem 2.24, we see that we can get an integer $h \geq H$ such that the values taken by the function $P_{m_h}(z)$ in the circle c cover one of the domains $\Omega_i(i = 1, 2)$, say Ω_1. So in c there is a point z' such that

$$P_{m_h}(z') = \alpha_1 . \tag{2.96}$$

Then by (2.94)

$$\alpha_1 \in J(P_{m_h}), \quad z' \in J(P_{m_h}) = J(P)$$

and by (2.93), (2.95), (2.96), we have

$$z' \neq z_0 .$$

Hence the point z' satisfies the condition (2.91).

2) λ is infinite. Let H be a positive integer such that

$$|P_{m_h}(z_0)| > M \text{ for } h \geq H \tag{2.97}$$

where M is the number in (2.92). As above, we can get an integer $h \geq H$ such that the values taken by the function $P_{m_h}(z)$ in the circle c cover one of the domains $\Omega_i(i = 1, 2)$. By (2.97), again we see that in c there is a point z' satisfying the condition (2.91).

It remains to prove the part 2° of Theorem 2.30. Let $a_j(j = 1, 2, \ldots, q)$ be all the distinct roots of the equation $P'(z) = 0$. Consider a point $z_0 \in J(P)$. Since z_0 is a point of accumulation of $J(P)$, for any circle $c : |z - z_0| < r$, there is a point $z' \in c$ such that

$$z' \in J(P), \quad z' \neq z_0, \quad z' \neq P(a_j) \ (j = 1, 2, \ldots, q) .$$

Let α be a root of the equation $P(z) = z'$. Then

$$P(\alpha) = z', \quad P'(\alpha) \neq 0 \ .$$

Next as in the proof of the part $3°$ of Theorem 2.24, we see that there is a point $z_1 \in c$ such that

$$z_1 \neq z_0, \quad z_1 \in E(P) \ .$$

Hence z_0 is a point of accumulation of $E(P)$.

For further study of the Julia sets of polynomials and rational functions, the reader is referred to the following works:

1. H. Brolin, "Invariant sets under iteration of rational functions", *Ark. Mat.* **6** (1965) 103-144.

2. P. Blanchard, "Complex analytic dynamics on the Riemann sphere", *Bull. American Math. Soc.* **11** (1984) 85-141.

3

FACTORIZATION OF MEROMORPHIC
FUNCTIONS

3.1. INTRODUCTION

The factorization theory of meromorphic functions concerns whether a given function can be expressed as a composition of two or more nonlinear meromorphic functions. This theory was developed about two decades ago. The investigation is closely related to the study of the fix-points of a function. A complex number z_0 is said to be a fix-point of $f(z)$ iff $f(z_0) = z_0$. As far back as 1926, Fatou claimed that for any nonlinear entire function f, the iteration $f(f)(= f_2)$ has at least one fix-point. This fact was formally proved in 1952 by P.C. Rosenbloom utilizing Picard's theorem. The proof assumes if $f(f)$ has no fix-point at all, then clearly, it is not possible for $f(z)$ to have any fix-point either. Therefore, the function

$$F(z) = \frac{f(z) - z}{f(f(z)) - f(z)}$$

is entire and assumes neither 0 nor 1. According to Picard's theorem F must be a constant, say c. Clearly $c \neq 1$. From the above equation, we get

$$c[f(f(z)) - z] = f(z) - z \ .$$

By differentiating both sides, we have

$$c[f'(f(z))f'(z) - 1] = f'(z) - 1$$

or

$$f'(z)[cf'(f(z)) - 1] = c - 1 \neq 0 .$$

It follows that $f'(z)$ never vanishes, and moreover, $f'(z)$ never takes the value $\frac{1}{c}$. Thus f' has to be a constant, hence f is linear. This contradicts the assumption and therefore proves the assertion that $f(f)$ must have infinitely many fix-points for any nonlinear entire function f.

In the same paper, Rosenbloom extended this result and obtained two theorems (given below) by applying the newly developed Nevanlinna's value-distribution theory. Since then, the value-distribution theory has greatly affected the research in the factorization theory.

Theorem 3.1. Let f and g be two transcendental entire functions. Then either f or $f(g)$ must have infinitely many fix-points.

Theorem 3.2. Let $P(z)$ be a nonlinear polynomial and $f(z)$ be a transcendental entire function. Then $P(f(z))$ must have infinitely many fix-points.

It was in the same paper that Rosenbloom first introduced the concept of "prime function". He defined an entire function $F(z)$ to be prime if every factorization of the form $F(z) = f(g(z))(= f \circ g(z))$. f, g being entire implies that one of the functions, f or g, must be linear. Rosenbloom asserted, without giving a proof, that $e^z + z$ is a prime function and remarked that its proof was quite complicated. Given the present techniques in the study of factorization, the proof that $e^z + z$ is prime is a relatively simple matter. It was not until 1968 that F. Gross gave a broader definition of the factorization for meromorphic functions. He not only provided a proof of the primeness of $e^z + z$ but also started a series studies on factorization theory. In this book, the emphasis will be on the development of the methods for testing whether a given meromorphic function can be factorized as two or more nonbilinear meromorphic functions. More specifically, we shall discuss

 (i) the forms of the factors in a factorization;
 (ii) the existence of fix-points and the factorization;
 (iii) the criteria for pseudo-prime functions;
 (iv) the growth rates of meromorphic solutions of certain functional equations and
 (v) the factorizations of meromorphic solutions of linear differential equations and the uniquely factorizability of certain functions.

Through the investigations of Gross and Yang in the U.S.A., Goldberg and Prokopovich in the U.S.S.R., Baker and Goldstein in England, Steinmetz in Germany, and Ozawa, Urabe and Noda in Japan; the theory of factorization has become a new branch in the value-distribution theory of meromorphic function. For the interest of the reader, we have included many open questions and research problems for further studies in this book.

3.2. BASIC CONCEPTS AND DEFINITIONS

Definition 3.1. Let $F(z)$ be a meromorphic function. If $F(z)$ can be expressed as

$$F(z) = f(g(z))(= f \circ g(z)) , \qquad (3.1)$$

where f is meromorphic and g is entire (g may be meromorphic when f is rational), then we call expression (3.1) a factorization of F (or simply a factorization), and f and g are called the left and right factors of F, respectively.

Definition 3.2. If every factorization of F of the above form, implies that either f or g is bilinear (f is rational or g is a polynomial), then F is called prime (pseudo-prime).

Definition 3.3. If every factorization of the form (3.1) leads to the conclusion that f must be a bilinear form when g is transcendental (or f is transcendental and g must be linear) then F is called left-prime (right-prime).

When factors are restricted to entire functions, it is called a factorization in entire sense. Under such a provision a prime (pseudo-) function will be denoted as E-prime (E-pseudo-prime). Nevertheless, we shall prove in the sequel that if a nonperiodic entire function F is E-prime (E-pseudo-prime) then it also must be prime (pseudo-prime). In other words, we only need to consider entire factors for the primeness (or pseudo-primeness) of a nonperiodic entire function.

Until recently the majority of the research accomplishments in the factorization theory have been based on the studies of the prototype $e^z + z$; the construction of certain families of prime or pseudo-prime functions; the finding of sufficient conditions for a certain class of functions being prime or pseudo-prime, and the discussions of problems of the uniquely factorizability of certain functions as well as the commutativity of factors. In all these investigations, the Nevanlinna value-distribution theory has been used as the primary tool. Also, in the development of the proofs, the following properties of meromorphic functions have been used: (i) the growth property; (ii) the distribution of the zeros or the existence of defect values; (iii) the periodicity; (iv) the fix-points and (v) being a solution of a linear differential equation. Generally speaking, the research in the factorization theory is still in its infancy stage. There are many interesting questions to be studied and resolved. We strongly believe that value-distribution theory can be further perfected by studying the factorization theory. Here we shall only deal with factorization of transcendental meromorphic functions, since Ritt has obtained a complete theory on the factorization of polynomials (but not rational functions!).

3.3. FACTORIZATION OF CERTAIN FUNCTIONS

In this section we shall prove that for any non-constant polynomial $P(z)e^z + P(z)$ is prime; a generalized form of $e^z + z$. Prior to this proof it is natural for us to ask: For a given function, is there any link between the forms of the factors and that of the given function? More precisely, we ask whether certain classes of entire functions and their factors possess more or less similar properties? The answer is "yes".

Before proceeding further, we introduce some definitions and lemmas below:

Definition 3.4. We shall call an entire function $F(z)$ periodic mod g with period τ, if and only if the following identity holds:

$$F(z + \tau) - F(z) = g(z) \ .$$

Sometimes, we simply call such a function a pseudo-periodic function mod g.

For instance given the function $F(z) \equiv e^z + P(z)$, F is periodic mod a polynomial with period $2\pi i$.

Theorem 3.3. Let $F(z)$ be an entire function and periodic mod a polynomial $P(z)$ with period τ. If $F = f \circ g$, then g must assume the following form:

$$g(z) = H_1(z) + q(z)e^{H_2(z)+cz} \tag{3.2}$$

where $H_i(z), i = 1, 2$, are periodic functions with the same period τ, c is a constant, and $q(z)$ is a polynomial.

Proof. We may assume, without loss of generality (w.l.o.g) that $\tau = 1$. According to the above hypothesis we have

$$f(g(z+1)) - f(g(z)) = P(z) .$$

We note that whenever $g(z_0 + 1) = g(z_0)$ for some z_0, the function on the left side of the above equation will assume a value zero. Therefore, we must have

$$g(z+1) - g(z) = P_1(z)e^{\alpha_1(z)} , \tag{3.3}$$

where $P_1(z)$ is a polynomial and $\alpha_1(z)$ is an entire function. Similarly, we deduce

$$g(z+2) - g(z) = P_2(z)e^{\alpha_2(z)} \tag{3.4}$$

where $P_2(z)$ is a polynomial and $\alpha_2(z)$ is an entire function. Substituting z with $z+1$ in equation (3.3) and then adding to equation (3.4) we get

$$g(z+2) - g(z) = P_1(z+1)e^{\alpha_1(z+1)} + P_1(z)e^{\alpha_1(z)} . \tag{3.5}$$

Equating (3.4) and (3.5) we get

$$P_1(z+1)e^{\alpha_1(z+1)} + P_1(z)e^{\alpha_1(z)} = P_2(z)e^{\alpha_2(z)} .$$

Applying Borel's lemma, we conclude that

$$\alpha_1(z+1) = \alpha_1(z) + c ,$$

c a constant. We write

$$H_2(z) = \alpha_1(z) - cz .$$

Then

$$H_2(z+1) - H_2(z) = \alpha_1(z+1) - cz - c = \alpha_1(z) - cz = H_2(z) .$$

Thus $H_2(z)$ is a periodic function with period 1. We easily verify (for instance, by equating the coefficients) that for any given polynomial $P_1(z)$ there always exists a polynomial $q(z)$ such that

$$e^c q(z+1) - q(z) = P_1(z) .$$

Setting

$$H_1(z) = g(z) - q(z)e^{\alpha_1(z)} = g(z) - q(z)e^{H_2(z)+cz} , \qquad (3.6)$$

we obtain

$$\begin{aligned}
H_1(z+1) &= g(z+1) - P_1(z+1)e^{H_2(z+1)+cz+c} \\
&= g(z) + q(z)e^{H_2(z)+cz} - e^c q(z+1)e^{H_2(z)+cz} \\
&= g(z) - (e^c q(z+1) - q(z))e^{H_2(z)+cz} \\
&= g(z) - P_1(z)e^{H_2(z)+cz} = H_1(z) .
\end{aligned}$$

Hence $H_1(z)$ is also a periodic function with period 1. It follows from (3.6) that

$$g(z) = H_1(z) + q(z)e^{H_2(z)+cz} ,$$

and the theorem is thus proved.

Remark. (i) If F is periodic mod a nonconstant entire function $h(z)$ with $\rho(h) \leq 1$, then the theorem remains valid, where $q(z)$ need not be a polynomial, but $\rho(q) \leq 1$.

(ii) In general, it seems there is not much that we can really say about the factors f and g in a factorization $F = f(g)$. However, for functions F of certain forms, we can determine the possible forms of f or g in the factorization $F = f(g)$.

Gross, Koont and Yang proved: Given $F(z) \equiv H_1(z) + ze^{H_2(z)}$, where H_1 and H_2 are periodic entire functions with the same period τ. For any factorization $F(z) = f(g(z))$, where f, g are entire functions with f being nonlinear, then $g(z)$ must be of the form $g(z) = T(z) + az$, where $T(z)$ is a periodic entire function with period τ and a is a non-zero constant. Furthermore they also proved if H_2 is prime, so is F.

Theorem 3.4. If $P(z)$ is any nonlinear polynomial and $g(z)$ is an arbitrary transcendental entire function, then $P(g)$ is not periodic mod a non-constant polynomial.

Proof. Assume the theorem to be false; then there exists a non-constant polynomial $l(z)$ such that

$$P(g(z+1)) - P(g(z)) = l(z) . \qquad (3.7)$$

By the above theorem, g is of the form

$$g(z) = H_1(z) + q(z)e^{H_2(z)+cz}$$

where $H_i(z), i = 1, 2$ are periodic entire functions with period 1, $q(z)$ is a polynomial, and c is a constant.

Substituting z by $z + 1, z + 2, \ldots, z + n - 1$ successively in (3.7) and adding them up, we obtain

$$P(g(z + n)) - P(g(z)) = l(z) + l(z + 1) + \ldots + l(z + n - 1) . \qquad (3.8)$$

Alternatively, from (3.7) we can derive

$$g(z + n) = g(z) + [q(z + n)e^{cn} - q(z)]e^{H_2(z) + cz} .$$

Substituting this into (3.8), we get

$$P(g(z) + [q(z + n)e^{cn} - q(z)]e^{H_2(z) + cz}) = P(g(z)) + l(z) + \ldots + l(z + n - 1).$$
$$(3.9)$$

If $|e^c| < 1$, then at $z = 0$, the right side of (3.9) tends to infinity with n, whereas the left side is bounded. This is a contradiction. When $|e^c| > 1$, replacing z by $-z$ in (3.9), we will arrive at a similar contradiction. Hence we may assume that $|e^c| = 1$. Let t be the degree of $l(z)$. When $z = 0$ and n is sufficiently large, then the absolute value of the right side of (3.9) is greater than

$$\lambda_1 \frac{n}{2} \left(\frac{n}{2}\right)^t$$

but less than

$$\lambda_2 (n - 1)^t (n - 1) ,$$

where λ_1, λ_2 are suitable constants. Assume the degree of $P = u$ and the degree of $q = v$ then from (3.9) we conclude that $uv = t + 1$. Now we shall treat the two cases separately: (a) $\alpha(z) = e^{H_2(z) + cz} \equiv \alpha$, a constant and (b) $\alpha(z)$ is not a constant. If case (a) holds, then

$$P(H_1(z) + \alpha q(z + 1)) - P(H_1(z) + \alpha q(z)) = l(z) .$$

For large $r(= |z|)$ the maximum modulus of the left side of the equation is greater than $\lambda (M(r, H_1))^{u-1}$ for some positive constant λ. This is impossible, since $u \geq 2$ and $H_1(z)$ is transcendental.

Now assume that case (b) holds. Then it can be easily verified that for every z the right side of (3.9) is less in absolute value than λn^{t+1} for some positive constant λ (independent of z) and sufficiently large $n > N(z)$; $N(z)$ is a quantity that depends on z. In order to estimate the left side of

(3.9), we may assume without loss of generality that $P(z)$ and $q(z)$ assume respectively the following forms:

$$P(z) = \lambda_u z^u + \dots \qquad (\lambda_u \neq 0)$$
$$q(z) = z^v + \dots .$$

We can choose z_0 so that

$$|\lambda_u||\alpha(z_0)|^u > 2\lambda . \tag{3.10}$$

Then for any $0 < \varepsilon < 1$ and sufficiently large n (depending on $N(z_0)$ and ε), the left side of (3.9) is greater in absolute value than

$$(1 - \varepsilon)|\lambda_u||\alpha(z_0)|^u n^{uv} > (1 - \varepsilon)2\lambda n^{t+1} . \tag{3.11}$$

It follows from this and (3.10) that

$$\lambda n^{t+1} > (1 - \varepsilon)2\lambda n^{t+1} ,$$

for sufficiently large n and any small ε. This will lead to a contradiction and the theorem is thus proved.

Remark. The theorem remains valid if the order of g is assumed to be less than 1, then the function $q(z)$ in the expression of $g(z)$ will satisfy $\rho(q) \leq 1$.

Theorem 3.5. If $P(z)$ is a polynomial of degree > 2 and if f is a transcendental entire function, then $f(P)$ is not periodic mod q (q a polynomial).

Proof. Suppose that the theorem is not true. We may then assume that the function $F(z) \equiv f(P(z))$ has a pseudo-period i such that

$$F(z + i) - F(z) = \varphi(z)$$

for some polynomial $\varphi(z)$ (φ may be a constant). It is known for any given polynomial $\varphi(z)$ that there exists a polynomial $q(z)$ such that

$$q(z + 1) - q(z) = \varphi(z) .$$

Hence $F(z) - q(z) \equiv H(z)$ is a periodic entire function with period i. Since f is transcendental, $F(z)$ must be transcendental. Thus for large r,

$$M(r, F) > r^2 M(2r, \varphi) . \tag{3.12}$$

Also for any positive integer n,

$$F(z + in) - F(z) = \varphi(z) + \varphi(z + i) + \ldots + \varphi(z + (n-1)i) . \qquad (3.13)$$

From the above two equations we can derive for every $\varepsilon > 0$, given sufficiently large $r > R(\varepsilon)$ and $|n| < r$:

$$(1 - \varepsilon)M(r, F) < M(r, F(z + ni)) < (1 + \varepsilon)M(r, F) .$$

Let $z_0 (|z_0| = r)$ be a point such that $|F(z_0)| = M(r, F)$. Recalling the degree of $P(z) \geq 3$, we can easily see that, for sufficiently large r and sufficiently small δ, there exists a point $z'(\neq z_0)$ in the sector $\delta < \arg z < \pi - \delta$ such that

$$\frac{|z'|}{|z_0|} \sim 1 \quad \text{and} \quad P(z_0) = P(z') .$$

Thus for $n > 0$,

$$M(r, F) = |f(P(z'))| = |F(z')|$$
$$= |F(z' + ni) - \varphi(z') - \varphi(z' + i) - \ldots - \varphi(z' + (n-1)i)| .$$

Using simple geometric illustration we can choose n so that
$z' + ni = x + iy$ with $|y| < \frac{1}{2}$ (when $n < 0$, the form (3.13) will be changed slightly, but the proof is not affected).

Using inequality (3.12), we obtain from the above equation that

$$M(r, F) < M(x + 1, F) + 2rM(2r, \varphi) < M(x + 1, F) + \frac{2}{r}M(r, F) ,$$

or

$$\left(1 - \frac{2}{r}\right) M(r, F) < M(x + 1, F) = M(1 + |z'| \cos \delta, F)$$
$$< M(\lambda r, F) \qquad (3.14)$$

for some λ satisfying $\cos \delta < \lambda < 1$. On the other hand by the Hadamard Three-circles theorem (choosing $r_1 = 1, r_2 = \lambda r$, and $r_3 = r$) we have

$$M(\lambda r, F) \leq kM(r, F)^{1 - \log \frac{1}{\lambda} / \log r}, \quad k \text{ a positive constant.}$$

If follows from the above equation that for any given positive $\varepsilon < 1$ we have

$$M(\lambda r, F) < \varepsilon M(r, F) \qquad (3.15)$$

for sufficiently large r.

Therefore Eqs. (3.14) and (3.15) lead to

$$\left(1 - \frac{2}{r}\right) < \varepsilon \ .$$

Since this is impossible, the theorem is proven.

Remark. Whittaker proved that for any given entire function g and constant c, the functional equation

$$cf(z+1) - f(z) = g(z)$$

has an entire solution $f(z)$ satisfying $\rho(f) = \rho(g)$ if $c = 1$. When $c \neq 1$ and $\rho(g) < 1$, Gross observed that the above functional equation has an entire solution f with $\rho(f) \leq 1$. In addition he asserted that if $\rho(f(P)) \geq 1$, then $f(P)$ is not periodic mod any function of order less than one.

Using the above theorems we can derive the following theorem.

Theorem 3.6. Let F be an entire function of exponential type and periodic mod some non-constant polynomial. Then F is either prime or of the form

$$F(z) = f((z+c)^2) \ ,$$

where f is an entire function and c is a constant.

Before proceeding with the proof, we need two additional results which are interesting in their own right.

Lemma 3.1. Let F be a non-periodic transcendental entire function. Then F is prime if and only if it is E-prime.

Proof. Assume that F is E-prime but not prime. Then there must be a factorization of the form

$$F = f \circ g \ ,$$

where f is a meromorphic (not entire) function. If f is transcendental, then f must be of the form

$$f(\varsigma) = (\varsigma - \varsigma_0)^{-n} f_1(\varsigma) \ ,$$

where n is a positive integer and f_1 is an entire function. Then g must have the form,

$$g(z) = \varsigma_0 + \varsigma^{\alpha(z)}, \quad \text{where } \alpha \text{ an entire function} \ .$$

Thus

$$F(z) = f \circ g(z) = e^{-n\alpha(z)} f_1(\varsigma_0 + e^{\alpha(z)})$$
$$= [e^{-n\varsigma} f_1(\varsigma_0 + \varsigma)] \circ \alpha(z) .$$

Since the left factor, $e^{-n\varsigma} f(\varsigma_0 + \varsigma)$ is a nonlinear entire function, and assuming that F is E-prime, we conclude that $\alpha(z)$ must be linear. This implies that g is periodic and so is F. This contradicts the hypothesis that F is not a periodic function. When f is rational (but not a polynomial), and using a similar argument to the one used in the preceding case, the same contradiction will be reached. This completes the proof of the lemma.

Equipped with this lemma, for the problem of factorizing non-periodic functions such as $e^z + P(z)$ (where $P(z)$ is a non-constant polynomial) we shall only need consider the entire factors. To prove that a transcendental entire function F is E-prime a general approach will be (i) to prove that F is pseudo-prime, and (ii) to show that when $F = f \circ P$ or $F = P \circ f$ with P being a polynomial, P has to be linear.

Lemma 3.2. Suppose that F is an exponential-type entire function

$$F(z + \tau) - F(z) = P(z)e^{az} ,$$

where $P(z)$ is a polynomial and a is a constant. Then F must be pseudo-prime.

Proof. Assume that $F = f \circ g$ and that f, g both are transcendental entire. Then

$$f(z(z + \tau)) - f(g(z)) = P(z)e^{az} . \tag{3.16}$$

Thus all the zeros of $g(z + \tau) - g(z)$ are among the zeros of $P(z)$. Hence,

$$g(z + \tau) - g(z) = q(z)e^{cz} , \tag{3.17}$$

where q is a polynomial and c is a constant. It is easy to conclude from the growth rate of a composite function that g is at most of order 1 and of minimum type. Consequently if $q(z)$ is not identically zero, then the constant c in (3.17) has to be zero. In any case, there exists a positive integer n that is sufficiently large so that $g^{(n)}(z + \tau) - g^{(n)}(z) = 0$. If $g^{(n)}$ fails to be a constant, then it is a periodic function and has at least a growth of order 1 and of finite type. This is a contradiction. Since $g^{(n)}(z)$ has to be a constant, $g(z)$ is a polynomial. This also proves that F is pseudo-prime.

We can now go back to the proof of Theorem 3.6.

Proof of Theorem 3.6. By Lemma 3.2, if F is not prime then F has only two possible factorizations: (a) $F = f \circ P$ or (b) $F = P \circ f$; f is transcendental and p is a nonlinear entire polynomial. However, according to Theorem 3.4, the factorization of form (b) can be ruled out. Finally, according to Theorem 3.5, F can only be factorized as $F = f \circ P$ when P is a polynomial of degree two. The theorem is thus proved.

Using the preceding theorem, we can immediately obtain the next one that was mentioned in the beginning of this chapter.

Theorem 3.7. Let $P(z)$ be any non-constant polynomial, and let $a(\neq 0), b$ be any two constants. Then $F(z) = e^{az+b} + P(z)$ is prime.

Proof. Using the last theorem, we only need to show that it is impossible to express $e^{az+b} + P(z)$ as $f((z+c)^2)$; f entire and c a constant. The proof of this simple exercise is left to the reader.

Remark. If in the theorem it is assumed that P is an entire function of order less than 1, then the conclusion remains valid.

Bascially we have proved that $e^z + z$ is prime. Ozawa proved that $e^{e^z} + z$ is prime. Ozawa, Yang, Gross, and Urabe have independently proven that for any positive integer, $e_n(z) + z$ is prime; $e_n(z)$ is the n-th iteration of e^z, i.e., $e_n(z) = e_{n-1}(e^z)$. These naturally lead to the question: For any given non-constant periodic entire function H, must $e^H + z$ be prime? Or more generally, must $H(z) + z$ be prime? No definite answer has been found for the former question (as far as we know of). For the latter one (in general it is not true!) we have the following result.

Theorem 3.8. Let H be a periodic entire function of finite lower order, and a be a nonzero constant. Then $H(z) + az$ is prime.

Discussion. If H is of infinite lower order, will the above theorem remain true? The answer is no. However, if in theorem 3.7, the polynomial P is a constant, then clearly F becomes a periodic entire function and is not prime. Gross raised this question: Does there exist a periodic entire function that is also a prime function? With regards to this question, Ozawa first provided some solutions that are E-prime. Later on, Yang and Gross also obtained some general results that are related to the above question. Ozawa exhibited some E-prime periodic entire functions of order 1 and infinite. We state the following result that seems to be a most general one in resolving Gross' question.

Theorem 3.9. (Yang and Gross) Let β be an arbitrary non-constant entire function, and P be a non-constant polynomial. Then

$$F(z) = (e^z - 1)\exp(P(e^{-z}) + \beta(e^z))$$

is E-prime.

The reader is referred to Gross and Yang's paper "On prime periodic entire functions", *Math. Zeit* **174** (1980) 43-48, for the proof. Note that in the text of the above paper, only the E-primeness of F has been established. It remains an open question whether there exists a periodic entire function which is also prime.

3.4. FACTORIZATION OF FUNCTIONS IN COSINE OR EXPONENTIAL FORMS

In this section we explore the possible forms of the factors in the factorization of functions in cosine or exponential forms. We first state (without giving a proof) the following lemma:

Lemma 3.3. (Edrei). Let $g(z)$ be an entire function. If there exists an unbounded sequence of numbers $\{a_n\}$ such that all, except finitely many of the values n, are the roots of the equations:

$$g(z) = a_n, \quad n = 1, 2, \ldots,$$

and lie on the same straight line, then $g(z)$ must be a polynomial with its degree not greater than 2.

It was then conjectured by Ozawa that if there is a sequence $\{a_n\}$ such that $|a_n| \to \infty$ and that all but except a finite number of roots of $f(z) = a_n$ lie on p straight lines: $l_1, l_2, \ldots, l_p$, any two of them are not parallel to each other, then $f(z)$ reduces to a polynomial of degree at most $2p$. Consequently, J. Qiao proved this conjectured for $p = 2, 3$ and recently, F. Ren presented a proof for any positive integer $p \geq 4$. As a consequence of these, one can show immediately that if f and g are two finite order entire functions with all but a finite number of their roots lie on finite straight lines, then the product $f(z)g(z)$ is pseudo-prime. For instance, $\sin z \sin \sqrt{2}z \sin \sqrt{3}z$ are pseudo-prime.

Lemma 3.4. Let f be an entire function of order less than $\frac{1}{2}$, and g be an arbitrary entire function. A necessary and sufficient condition for $F = f \circ g$ being periodic is that g is a periodic function.

Proof. Assume that F has a period τ. Choose a point z_0 and avoid $\alpha \; (= F(z_0))$ a branch point of the inverse function f^{-1}. Consider the straight line $L : \varsigma = z_0 + t\tau, -\infty < t < \infty$. Then clearly F is bounded on L. We are going to show that g is also bounded on L. If g is unbounded on L, then $g(L)$, the image of L under g, is a path tending to infinity on which f is bounded. Since the order of f is $< \frac{1}{2}$, by Wiman's theorem this is impossible. Therefore, $\{g(z_0 + n\tau)\}_{n=1}$ is bounded. Hence, there exists a constant $k > 0$ such that

$$|g(z_0 + n\tau)| \leq k \quad \forall n = 1, 2, \ldots .$$

We also have for $n = 1, 2, \ldots$

$$f(g(z_0 + n\tau)) = f(g(z_0)) = \alpha .$$

Thus all the points in $\eta_n = g(z_0 + n\tau)$ are among the finite roots of the equation $f(\eta) = \alpha$ that lie in $|\eta| \leq k$. Hence, there must exist some $m \neq n$ such that $g(z_0 + m\tau) = g(z_0 + n\tau)$. Moreover, for sufficiently small ε,

$$\begin{aligned} f(g(z_0 + \varepsilon + m\tau)) = F(z_0 + \varepsilon + m\tau) &= F(z_0 + \varepsilon + n\tau) \\ &= f(g(z_0 + \varepsilon + n\tau)) . \end{aligned}$$

Now since f is one-one in a sufficiently small region around the point $g(z_0 + \varepsilon + n\tau)$ (by choosing $\alpha = f(g(z_0)))$, we must have $g(z_0 + \varepsilon + m\tau) = g(z_0 + \varepsilon + n\tau)$ for all sufficiently small ε. Hence $g(z_0 + m\tau) \equiv g(z_0 + n\tau)$, and g has period $(m - n)\tau$. This completes the proof of the lemma.

Theorem 3.10. The function $F(z) = \cos z$ is pseudo-prime. Furthermore, the possible forms of the factorization of $F = f \circ g$ are as follows:

(i) $f(\varsigma) = \cos \sqrt{\varsigma}, g(z) = z^2$;

(ii) $f(\varsigma) = T_n(\varsigma), g(z) = \cos \frac{z}{n}$,

where $T_n(\varsigma)$ denotes the nth Chebyshev polynomial $(n \geq 2)$,

(iii) $f(\varsigma) = \frac{1}{2}(\varsigma^{-n} + \varsigma^n), g(z) = e^{iz/n}; n$ denotes a non-negative integer.

Proof. The E-pseudo-primeness of $\cos z$ follows immediately from Lemma 3.2. We now show that $\cos z$ is also pseudo-prime. Assume that $\cos z = f \circ (g(z))$, where $f(\varsigma)$ is a transcendental (non-entire) function, and g is transcendental entire. Then by Pólya's theorem $\rho(f) = 0$, and, moreover, f has at most one pole. Hence, $f(\varsigma) = (\varsigma - \varsigma_0)^{-n} f_1(\varsigma)$ for $n \geq 1$; f_1 is an entire function with $f_1(\varsigma_0) \neq 0$, and $g(z) = \varsigma_0 + e^{M(z)}$; M is a non-constant entire function. Thus

$$\cos z = f_1(g(z))e^{-nM(z)} .$$

Let $\{\eta_j\}$ be the zero set of $f_1(\varsigma)$ that is an unbounded set. To each η_j, all the roots of the equation $g(z) = \eta_j$ are real since $\cos z$ has only real zeros. By Lemma 3.2, we conclude that $g(z)$ is a polynomial of degree no greater than 2. This is a contradiction. Hence, $\cos z$ is pseudo-prime. Now we investigate the cases where the factors are restricted to entire functions only.

(a) $\cos z = f \circ g(z)$; f entire, g a polynomial. By the above argument, it follows that g is a polynomial of degree no greater than 2. Then g can be expressed as

$$g(z) = k(z - a)^2 + b; \quad k \neq 0, \quad a, b \text{ are all constants} .$$

This is the form (i) mentioned in Theorem 3.10.

(b) $\cos z = P_n \circ g(z)$, where P_n is a polynomial of degree ≥ 2, and g is an entire function. By Lemma 3.4, we conclude that g is a periodic function with period, say τ. We have

$$\cos(z + \tau) = P_n(g(z + \tau)) = P_n(g(z)) = \cos z , \qquad (3.18)$$

then, $\tau = 2l\pi$ for some integer $l(\neq 0)$ and $g(z)$ is a function of an exponential type and is periodic with period $2l\pi$. It can be expressed as

$$g(z) = \sum_{k=-m}^{m} a_k e^{ikz/l} .$$

Substituting this into (3.18) and applying Borel's lemma, we can conclude $l = n, m = 1$. Then $g(z)$ can be written as

$$g(z) = a_0 + a_1 e^{iz/n} + a_{-1} e^{-iz/n} .$$

We may assume without loss of generality that $a_0 = 0, a_1 = \frac{1}{2}$, and can set

$$P_n(\eta) = A_n\eta^n + A_{n-1}\eta^{n-1} + \ldots + A_1\eta + A_0 .$$

Using the above two expressions and (3.18), we have

$$\frac{1}{2}e^{iz} + \frac{1}{2}e^{-iz} = P_n\left(\frac{1}{2}e^{iz/n} + a_{-1}e^{-iz/n}\right) . \qquad (3.19)$$

Equating the coefficients of e^{iz} and e^{-iz} respectively on both sides of (3.19), we get $A_n = 2^{n-1}$ and $a_{-1}^n = 2^{-n}$. Hence,

$$a_{-1} = e^{2s\pi i/n}, \quad s = 0, 1, 2\ldots, n-1 .$$

And consequently,

$$g(z) = g_{n,s}(z) = \frac{1}{2}e^{iz/n} + \frac{1}{2}e^{2s\pi i/n}e^{-iz/n}$$

$$= e^{s\pi i/n} \cos\frac{z - s\pi}{n} , \quad (s = 0, 1, 2, \ldots, n-1) .$$

Let $T_n(w)$ denote the n-th Chebyshev polynomial, namely

$$T_n(\cos z) = \cos nz . \qquad (3.20)$$

Equations (3.19) and (3.20) give us,

$$P_n(\eta) = P_{n,s,}(\eta) = (-1)^s T_n(e^{-s\pi i/n}\eta), \quad s = 0, 1, 2, \ldots, n-1 .$$

In fact,

$$P_{n,s}(g_{n,s}(z)) = (-1)^s T_n(e^{-s\pi i/n}g_{n,s}(z)) = (-1)^s T_n\left(\cos\frac{z - \pi s}{n}\right) .$$

Setting $s = 0$, we get $g(z) = \cos z/n$ and $P_n(\eta) = T_n(\eta)$. Thus, to a fixed n, all the factorizations of $P_{n,s} \circ (g_{n,s})(s = 1, 2, \ldots n-1)$ are equivalent to $P_{n,0} \circ g_{n,0}$, that is the desired form (ii) of the theorem.

(c) Finally, let us assume $\cos z = f \circ g(z)$, where f is rational (but not a polynomial) and g is transcendental entire. Then as before we have,

$$f(\varsigma) = (\varsigma - \varsigma_0)^{-n}P(\varsigma), \quad g(z) = \varsigma_0 + e^{\alpha(z)} ,$$

where $n \geq 1$, P is a polynomial of degree t, and $\alpha(z)$ is a non-constant entire function. Obviously the function α has to be linear, namely $\alpha(z) = az + b; a \neq 0$ and b is a constant. Thus

$$\cos z = P(\varsigma + e^{az+b}) \exp\{-n(az+b)\} \ .$$

We may, without loss of generality, assume that $\varsigma_0 = 0$. It follows from the above two expressions that

$$\cos z = a_t e^{(t-n)az} + a_{t-1} e^{(t-n-1)az} + \ldots + a_0 e^{-naz} \ ,$$

where $a_t, a_{t-1}, \ldots, a_0$ are suitable constants; $a_t \neq 0$. By an application of Borel's lemma, we immediately obtain

$$t = 2n, \quad a_{t-1} = \ldots = a_1 = 0 \ ,$$

and

$$a = \frac{i}{n}, \quad a_t = \frac{1}{2}, \quad a_0 = -\frac{1}{2} \ ,$$

or

$$a = \frac{i}{n}, \quad a_t = -\frac{1}{2}, \quad a_0 = -\frac{1}{2} \ .$$

Form (iii) is now established, and the theorem is proven.

Remark. Another justification that we shall learn about in the sequel (Chap. 4) for $\cos z$ being pseudo-prime is that it satisfies an ordinary linear differential equation: $y''(z) - y(z) = 0$.

Theorem 3.10 shows that for any given positive integer k, $\cos z$ can be factorized as

$$\cos z = P_k \circ g_k(z) \ , \tag{3.21}$$

where P_k is a suitable polynomial of degree k, and g_k is a corresponding entire function.

Ozawa established a converse to Theorem 3.10 as follows.

Theorem 3.11. Let F be a non-constant entire function. Suppose that for each integer $k = 2^j (j = 1, 2, \ldots)$ and $k = 3$ or 5,

$$F(z) = P_k \circ g_k(z) \tag{3.22}$$

holds for some polynomials P_k and entire function g_k. Then either

$$F(z) = a e^{H(z)} + b \tag{3.23}$$

or

$$F(z) = a \cos \sqrt{H(z)} + b , \tag{3.24}$$

where $a(\neq 0)$ and b are constants and $H(z)$ is an entire function.

Remark. Ozawa also proved that the conclusions made in Theorem 3.11 remain valid under weaker hypotheses; namely (3.22) holds for $k = 3^j (j = 1, 2, \dots)$ and $k = 2, 4$.

Ozawa's proof is rather complicated. Recently, J. Huang and G.D. Song improved Ozawa's method and obtained the following stronger result with a simplified argument.

Theorem 3.12. Let $F(z)$ be a non-constant entire function. Suppose that (3.22) holds for each integer $k = q$ (a given positive integer) $k = n_j$ with $2 \leq n_1 < q, (n_j, q) = 1, n_{j+1} \leq n_j q, j = 1, 2, \dots$ .Then $F(z)$ has either form (3.23) or form (3.24).

Lemma 3.5. Let $Q(\varsigma)$ be a given polynomial of degree p. Then there exist only finitely many pairs of linear functions U and V such that

$$V \circ Q = Q \circ U ,$$

unless $Q(\varsigma) = A(\varsigma - \alpha)^p + B, A, B$ are constants.

Proof. Let $V(t) = at + b, U(t) = ct + d$. If

$$V \circ Q = aQ(\varsigma) + b = Q \circ U = Q(c\varsigma + d) ,$$

then, by differentiating, we have

$$aQ'(\varsigma) = cQ'(c\varsigma + d) .$$

This shows that the zero set Z of Q' is invariant under mapping: $\varsigma \to c\varsigma + d$. Moreover $c \neq 1$ (since Z is a finite set, unless $V(\varsigma) = U(\varsigma) = \varsigma$). Then by setting $\alpha = d/(1 - c)$, we have

$$U(t) = c(t - \alpha) + \alpha .$$

Therefore, unless $Z = \{\alpha\}$, the invariant property of Z leads to the conclusion that $|c| = 1$. If $Z \neq \{\alpha\}$, then c is a root of unity, and so we may assume $c^N = 1$, but $c^k \neq 1 (0 < k < N)$. Thus Z consists of the vertices of a regular N-polygon centered at the point α and possibly the point α itself. It follows that

$$Q'(\varsigma) = c(\varsigma - \alpha)^s \prod_{j=1}^{M} \{(\varsigma - \alpha)^N - b_j\} \ .$$

By integrating, we get

$$Q(\varsigma) = (\varsigma - \alpha)^{s+1} h[(\varsigma - \alpha)^N] + B \ ,$$

where h is a polynomial and B is a constant. In the meantime, we also have

$$Q \circ U(\varsigma) = c^{s+1} Q(\varsigma) + B(1 - c^{s+1}) = V \circ Q(\varsigma) \ ,$$

and thus

$$V(t) = c^{s+1} t + B(1 - c^{s+1}), \quad (c^N = 1) \ .$$

It is easy to see from the above two expressions that there can be only finitely many such $V(t)$, so the number of the function $U(t)$ is also finite. Now, suppose that $Z = \{\alpha\}$, then

$$Q'(\varsigma) = c(\varsigma - \alpha)^{p-1}, \quad Q(\varsigma) = A(\varsigma - \alpha)^p + B \ .$$

Under this circumstance, it is obvious that there are infinitely many pairs of U, V with $U(t) = c(t - \alpha) + \alpha, V(t) = c^p t + B(1 - c^t)$ and satisfying $V \circ Q = Q \circ U$. Therefore Lemma 3.5 is proven.

Lemma 3.6. Let $f(s)$ and $g(z)$ be two non-constant entire functions. Let P_m and P_n be two polynomials of relatively prime degrees m and n, respectively. If for all z which belong to the complex plane the following identity holds:

$$P_m \circ f(z) = P_n \circ g(z) \ ,$$

then there exist entire function $s(z)$, and polynomials $U(\varsigma)$ and $V(\varsigma)$ of degree n and m, respectively such that

$$f(z) = U(s(z)), \quad g(z) = V(s(z)) \ .$$

Proof. We first factorize $P_m(u) - P_n(v)$ into irreducible factors in $C[u, v]$. Let $R(u, v)$ be such a factor, then we have

$$R(f(z), g(z)) = 0 \ .$$

Using a theorem of Picard's that can be phrased as: if $R(u, v)$ is an irreducible polynomial in $C[u, v]$ and there exist some non-constant entire functions $f(z)$ and $g(z)$ satisfying $R(f(z), g(z)) = 0$, for all z, then the Riemann surface of $R(u, v) = 0$ has genus 0 and will be denoted by S (conformationally equivalent to the Riemann sphere). If f and g in $R(f(z), g(z)) = 0$ can be meromorphic then the genus of such a surface X is less than or equal to 1. Moreover, any Riemann surface of genus 1 can only be uniformized by elliptic functions not by entire functions, and there exists a conformal map ψ that maps the points p on the Riemann surface R (defined by $R(u, v) = 0$) onto points ς of the Riemann sphere; i.e., $\varsigma = \psi(p)$. We may assume without loss of generality that $\varsigma = \infty$ corresponds to a point with $u = \infty$.

Except at a finite numbering branch points of R we may use u as a local uniform parameter. Thus, except at those branch points, ς is a holomorphic function $\sigma(u)$ of $u \in R$. Therefore, the map

$$z \to (f(z), g(z)) = p \to s = \psi(p) = \sigma \circ f(z) = s(z)$$

is holomorphic at all $z \in R$, except perhaps those for which $(f(z^*), g(z^*)) = (u^*, v^*) = p^*$ is a branch point of R. Clearly these exceptional points z^* form a discrete set E. Also if z tends to $z^* \in E$ this implies $s(z)$ tends to $s(z^*)$. Therefore, z^* is a removable singularity of $s(z)$, and hence $s(z)$ is entire.

Now the Riemann surface S, conformally equivalent to the Riemann sphere, is defined by $R(u, v) = 0$ and has genus zero. Then there exist uniformizable functions

$$u = U(\varsigma) , \quad v = V(\varsigma) ,$$

where U, V are rational functions such that a $1 - 1$ correspondence between points of S and the Riemann sphere is established. It follows that there exists $u = U(\varsigma)$ and $v = V(\varsigma)$ where U, V are rational functions on the surface R mentioned before, such that

$$f(z) = U(s(z)) \text{ and } g(z) = V(s(z)) . \tag{3.25}$$

Suppose $R(u, v)$ is of degree m_1 in u, n_1 in v. Any given value v, u has, in general, m_1 possible values (counting the multiplicities). That is, there are m_1 values of s for a given v. Therefore V is of degree m_1. In a similar manner, we can verify that U is of degree n_1. Since $f(z)$ and $g(z)$ are both entire, U and V can have one pole at finite value ς. Moreover, to any value z, $s(z) \neq \varsigma_0$, we may assume without loss of generality, that $\varsigma_0 = 0$ (otherwise replace ς by $\varsigma - \varsigma_0$). It follows from the above analysis on U and V that for some non-negative integers $s(\leq n_1)$ and $t(\leq m_1)$

$$U(\varsigma) = \sum_{k=-s}^{n_1-s} a_k \varsigma^k , \quad V(\varsigma) = \sum_{k=-t}^{m_1-t} b_k \varsigma^k . \tag{3.26}$$

It follows that $P_m \circ f(z) = P_n \circ g(z)$ with (3.25) yields

$$P_m \circ U \circ s(z) = P_n \circ V \circ s(z), \quad \forall z \in \mathbb{C} \text{ (the complex plane)} .$$

Since $s(z)$ assumes an infinite number of distinct values, the above equation leads to

$$P_m \circ U(\varsigma) = P_n \circ V(\varsigma) \quad \forall \varsigma \in \mathbb{C} .$$

From this result and (3.26), for sufficiently large values of ς, we can derive the following asymptotic relations:

$$P_m \circ U(\varsigma) \sim d_1 \varsigma^{(n_1-s)m} , \quad P_n \circ V(\varsigma) \sim d_2 \varsigma^{(m_1-t)n}$$

where d_1 and d_2 are two nonzero constants. Thus

$$(n_1 - s)m = (m_1 - t)n .$$

Recalling that $(m, n) = 1, 0 \leq n_1 - s \leq n, 0 \leq m_1 - t \leq m$ we can conclude that $n_1 - s = n_1$ and $m_1 - t = m$. Hence either $n_1 = n, s = 0$ and $m_1 = m, t = 0$ or $n_1 = s, m_1 = t$. This shows that either U and V are polynomials of degree n and m respectively or they are polynomials in $1/\varsigma$.

We then can make the substitution $W(z) = \frac{1}{s(z)}$ (noting that $s(z) \neq 0 \ \forall z \in C$), and proceed with the discussion on the polynomials, $U_1(w) = U(\varsigma)$ and $V_1(W) = V(\varsigma)$. We shall arrive at the same conclusion: $n_1 = n$ and $m_1 = n$. The lemma is proven.

J.F. Ritt made a further study on the equation:

$$P_m \circ U_n = P_n \circ V_m \tag{3.27}$$

where P_m, P_n, U_n and V_m are polynomials of degree m, n, n, m respectively; $(m, n) = 1$. He obtained the following result. ["Prime and composite polynomials," *Trans. Amer. Math. Soc.* **23** (1922) 51-66].

Lemma 3.7. The Eq. (3.27) can hold only in the following cases:

(A) There exists linear polynomials $\lambda_1, \lambda_2, \lambda_3$ and λ_4 such that

$$\lambda_1 \circ P_m \circ \lambda_2(u) = T_m, \quad \lambda_1 \circ P_n \circ \lambda_3 = T_n, \quad \lambda_2^{-1} \circ U_n \circ \lambda_4 = T_n,$$
$$\lambda_3^{-1} \circ V_m \circ \lambda_4 = T_m, \quad \lambda_1 \circ P_m(U_n) \circ \lambda_4 = \lambda_1 \circ P_n(V_m) \circ \lambda_4 = T_{nm} ,$$

where the T_k denotes kth Chebyshev polynomial.

(B) Suppose $m > n$. There exists linear polynomials $\lambda_1, \lambda_2, \lambda_3, \lambda_4$ and a polynomial h of degree less than m/n such that

$$\lambda_1 \circ P_m \circ \lambda_2(u) = u^2 h(u)^n \quad (r + n \deg h = m) ,$$
$$\lambda_2^{-1} U_n \circ \lambda_4(s) = s^n ,$$
$$\lambda_1 \circ P_n \circ \lambda_3(u) = U^n ,$$
$$\lambda_3^{-1} \circ V_m \circ \lambda_4(s) = s^r h(s^n) ,$$
$$\lambda_1 \circ P_m(U_n) \circ \lambda_4(s) = \lambda_1 \circ P_n(V_m) \circ \lambda_4(s) = [s^r h(s^n)]^n .$$

Proof. Without loss of generality we may assume,

$$1 < n_1 < q < n_2 < n_3 < \ldots ; n_{j+1} \le q n_j, j = 1, 2, \ldots \qquad (3.28)$$

Applying Lemma 3.5 with $m = n_j (j \ge 2), n = q$, we get polynomials U (of degree q), V (of degree m), and entire function $s_m(z)$ such that:

$$F(z) = P_m \circ U(s_m(z)) = P_q \circ V(s_m(z)) . \qquad (3.29)$$

Now according to Lemma 3.7, the polynomials $P_m, U, P_q,$ and V must satisfy one of the cases, (A) or (B). We shall show that if case (B) holds for a certain pair $(m = n_j, n = q)$, then (B) will hold for any other pairs (n_k, q). Alternatively suppose that we consider case (A) for a certain pair (n_k, q). There exists linear polynomials ρ and σ such that,

$$\rho \circ P_q \circ \sigma(u) = T_q(u) ,$$

where T_q is qth Chebyshev polynomial. On the other hand, case (B) holds for m and $n = q < m$. Thus there are linear polynomials μ, λ such that

$$\lambda \circ P_q \circ \mu(u) = V^q \circ u .$$

Hence

$$\rho \circ \lambda^{-1} \circ V^q \circ \mu^{-1} \circ \sigma(u) = T_q(u) ,$$

or

$$A(Bu + C)^q + D = T_q(u) .$$

Note that T_q does not assume any value of multiplicity ≥ 3. The above equation leads to a contradiction, since $q \geq 3$. Therefore, Theorem 3.12, will be a consequence of the results of the following two results, Theorems 3.13 and 3.14.

Theorem 3.13. If there exists an infinite sequence $M = \{m_k\}$ and integer $q \geq 3$ satisfying $(q, m_k) = 1$ (for $k = 1, 2, \ldots$) such that (3.29) and case (A) of Lemma 3.7 hold, then F must have the form: $F(z) = a\cos\sqrt{H(z)} + b$; where $a(\neq 0)$ and b are constants, and $H(z)$ is an entire function.

Theorem 3.14. If there exists a sequence of values of $m(= \{m_k\})$ and $n = q$ satisfying the condition of Eq. (3.28) such that Eq. (3.29) and case (B) in Lemma 3.7 hold, then F must have the form:

$$F(z) = ae^{H(z)} + b$$

where $a(\neq 0)$ and b are constants and $H(z)$ is an entire function.

Proof of Theorem 3.13. By case (A), corresponding to each $m \in M$, there exists linear polynomials λ_m and ν_m such that

$$\lambda_m \circ P_q \circ \nu_m = T_q .$$

If $\tilde{\lambda}_m$ and $\tilde{\nu}_m$ are the linear polynomials that correspond to $\tilde{m} \in M$. Then we have

$$\lambda_m \circ \tilde{\lambda}_m^{-1} \circ T_q \circ \tilde{\nu}_m^{-1} \circ \nu_m = T_q .$$

Recalling that for $q \geq 3$, $T_q(u)$ cannot be expressed as $A(u - \alpha)^q + B$, from this and Lemma 3.5, we can conclude that there can only be a finite number of pairs $(\lambda_m \circ \tilde{\lambda}_m^{-1}, \tilde{\nu}_m^{-1} \circ \nu_m)$. Keeping m fixed and replacing M by an infinite subsequence N, if necessary; we may assume that λ_m does not depend on the choice of m. It follows from Eq. (3.29) and (A) that there exists a linear polynomial λ such that

$$\lambda \circ F(z) = T_{q_m}(\tilde{s}_m(z)) , \quad \forall m \in N ,$$

where $\tilde{s}_m(z)$ is a composition of the function $s_m(z)$ in Eq. (3.29) and a linear polynomial.

Setting

$$\tilde{s}_m(z) = \cos \phi(z) \; , \tag{3.30}$$

we have

$$\lambda \circ F(z) = \cos(qm\phi(z)) \; . \tag{3.31}$$

The expression ϕ in Eq. (3.30) is a multivalued function but in a disk D (in the z-plane), that contains no roots of $\tilde{s}_m(z) = \pm 1$, we can define ϕ without ambiguity as a specific branch of $\cos^{-1} \tilde{s}_m(z)$. If ϕ_0 is one of the branches of $\cos^{-1} \tilde{s}_m(z)$, then any other branch $\phi(z)$ is determined by the formula:

$$\phi(z) = \pm\phi_0(z) \quad (\mathrm{mod} 2\pi), \quad \forall z \in D \; .$$

Let n be a member in N other than m. We can obtain functions $\tilde{s}_n(z)$ and $\psi(z)$ having properties similar to that of $\tilde{s}_m(z)$ and $\phi(z)$ respectively such that

$$\tilde{s}_n(z) = \cos \psi(z) \; , \tag{3.32}$$

$$\lambda \circ F(z) = \cos(qn\psi(z)) \; . \tag{3.33}$$

Similarly, for the multivalued function ψ, we can define in a disk (that incidentally can be chosen to be identical to D) a specific branch ψ_0 such that

$$\psi(z) = \pm\psi_0(z) \quad (\mathrm{mod} \; 2\pi), \quad \forall z \in D \; . \tag{3.34}$$

From Eqs. (3.31) and (3.33) we have

$$qn\psi_0(z) = \pm qm\phi_0(z) \quad (\mathrm{mod} \; 2\pi) \; .$$

We may suppose (if necessary, by changing ψ_0 into $-\psi_0$)

$$qn\psi_0(z) = qm\phi_0(z) + 2k\pi \; .$$

Hence,

$$\psi_0(z) = \frac{m}{n}\phi_0(z) + \frac{2k\pi}{qn} \; . \tag{3.35}$$

We set $\overline{\psi}_0(z) = \psi_0(z) + 2t\pi$, where t is suitable integer, such that

$$\overline{\psi}_0(z) = \frac{m}{n}\phi_0(z) + c; \quad c \text{ is a constant satisfying } -\pi < c \le \pi \; .$$

Both ϕ_0 and $\overline{\psi}_0$ can be analytically continued along any path Γ that avoids the roots of $\tilde{s}_m(z) = \pm 1$ and $\tilde{s}_n(z) = \pm 1$. Particularly, if Γ is chosen to be a closed path in D, then the continuation ϕ_Γ and $\overline{\psi}_\Gamma$ still satisfy Eqs. (3.30), (3.32), and (3.35). Let

$$\phi_\Gamma(z) = \delta\phi_0(z) + 2l\pi, \quad \delta = \pm 1, \text{ where } l \text{ is an integer} .$$

Then

$$\overline{\psi}_\Gamma(z) = \frac{m}{n}\phi_\Gamma(z) + c = \delta\frac{m}{n}\phi_0(z) + \frac{2ml\pi}{n} + c \qquad (3.36)$$

and also

$$\overline{\psi}_\Gamma(z) = \eta\overline{\psi}(z) + 2k\pi = \eta\frac{m}{n}\phi_0(z) + \eta c + 2k\pi; \quad \eta = \pm 1 .$$

Comparing the above two expressions and noting that k depends on n and Γ, l on m and Γ, and c on m and n only. We see that $\delta = \eta$. Therefore

$$c + 2lm\pi/n = \eta c + 2k\pi .$$

If $\eta = 1$, this leads to $l\frac{m}{n} = k$. Since n must be sufficiently large, we must conclude that k and hence l are both zero. If $\eta = -1$. We have $2k\pi = 2c + 2lm\pi/n$. By assuming $-\pi < c \leq \pi$ and for large n, we get $k = 0$ or -1. Since c is independent of the path Γ, when $\eta = -1$, k is also independent of Γ. Consequently the relationship between ψ_Γ and $\overline{\psi}_0$ can be summarized into the following cases.

    (i) $k = 0, \psi_\Gamma = \overline{\psi}_0(\eta = 1)$ or $\psi_\Gamma = -\overline{\psi}_0 \quad (\eta = -1)$

    (ii) $k = 1, \psi_\Gamma = -\overline{\psi}_0$ or $\psi_\Gamma = 2\pi - \overline{\psi}_0$

    (iii) $k = -1, \psi_\Gamma = \overline{\psi}_0$ or $\psi_\Gamma = 2\pi - \overline{\psi}_0$.

Thus we can see that $\psi^2, (\psi - \pi)^2$ and $(\psi + \pi)^2$ are single-valued analytic functions of z where $\tilde{s}_n(z) \neq \pm 1$. Now ψ is locally defined to be a branch of $\cos^{-1}(\tilde{s}_n(z))$ and approaches a finite value as z tends to a root of equations $\tilde{s}_n(z) = \pm 1$. Therefore, all these roots are removable singularities of the above three functions. Under case (i)

$$\psi^2(z) = H(z)/(qn)^2$$

is an entire function. Moreover, $\lambda \circ F(z) = \cos\sqrt{H(z)}$, and $F(z)$ assumes the form in Eq. (3.24). For the other two cases, we have

$$\psi(z) \pm \pi = \sqrt{H(z)}/qn .$$

Therefore, $\lambda \circ F(z) = \cos(\mp q n \pi + \sqrt{H(z)}) = \pm \cos \sqrt{H(z)}$. Thus $F(z)$ assumes the form in Eq. (3.24). The theorem is thus proved.

Proof of Theorem 3.14. Letting $m = n_k$ and $n = q$, in Eq. (3.29) and relation (B), we see that there exists a linear polynomial and entire function $s_k(z) = A_k s_{n_k}(z) + B_k$ such that

$$\lambda_k \circ F(z) = [s_k^{r_k} h_k(s_k^q)]^q \quad (k = n_2, n_3, \dots) , \tag{3.37}$$

where the h_k are polynomials with $h_k(0) \neq 0$, and $r_k + q \deg h_k = n_k$. We may assume without loss of generality that $h_k(0) = 1$. By applying the last form in relation B for the indices n_1 and q, there exists a linear polynomial λ_0 such that

$$\lambda_0 \circ F(z) = [s^n h(s^{n_1})]^{n_1} , \quad r + n_1 \deg h = q . \tag{3.38}$$

If α and β are roots of $\lambda_0(t) = 0$ and $\lambda_k(t) = 0$ respectively, we shall show that $\alpha = \beta$. Assume that $\alpha \neq \beta$, then by Nevanlinna's second fundamental theorem, we have

$$T(r, F) \leq \overline{N}(r, F) + \overline{N}\left(r, \frac{1}{F - \alpha}\right) + \overline{N}\left(r, \frac{1}{F - \beta}\right) + o\{T(r, F)\} \tag{3.39}$$

for a suitable sequence of r values tending to ∞.

On the other hand, we observe that each root of the equations, $F(z) = \alpha$ and $F(z) = \beta$ has multiplicity of at least n_1 and q, respectively. Hence

$$\overline{N}\left(r, \frac{1}{F - \alpha}\right) \leq \frac{1}{n_1} N\left(r, \frac{1}{F - \alpha}\right) \leq \frac{1}{n_1}(1 + o(1))T(r, F) ,$$

and

$$\overline{N}\left(r, \frac{1}{F - \beta}\right) \leq \frac{1}{q} N\left(r, \frac{1}{F - \beta}\right) \leq \frac{1}{q}(1 + o(1))T(r, F) .$$

Substituting the above two results into Eq. (3.39) and noting that $\overline{N}(r, F) = 0$, we get

$$T(r, F) \leq \left(\frac{1}{q} + \frac{1}{n_1} + o(1)\right) T(r, F) ,$$

which is impossible. It follows that $\alpha = \beta$, and hence $\lambda_k(t) = c \lambda_0(t)$.

Replacing $s_k(t)$ by $\delta s_k(t)$, if necessary, we may assume that $c = 1$, i.e.,

$$\lambda_0 = \lambda_k \quad (k = 2, 3, \dots) .$$

Thus

$$F_0(z) = \lambda_0 \circ F(z) = [s_k^{r_k} h_k(s_k^q)]^q = [s^r h(s^{n_1})]^{n_1} . \tag{3.40}$$

By assuming $(n_1, q) = 1$, we easily see from the above expression that the multiplicity of every root of $F_0(z) = 0$ is divisible by $n_1 q$. Hence, we have

$$F_0(z) = [G(z)]^{n_1 q} ,$$

where G is an entire function. From this and Eqs. (3.38) and (3.40), we have

$$G^q = s^r h(s^{n_1}) .$$

We let

$$h(t) = \prod_a (t - a^{n_1})^{\mu(a)} ,$$

where $\mu(a)$ denotes the multiplicity of the root a^{n_1}. Then

$$h(t^{n_1}) = \prod_a \prod_{j=1}^{n_1} (t - \rho^j a)^{\mu(a)}$$

where ρ is a primitive n_1-th root of unity.

If z_1 is a zero of $s(z) - \rho^j a$ with multiplicity ν, then $q|\nu\mu(a)$

Suppose that $q \nmid \mu(a)$ then $\nu \geq 2$, and $\rho^j a$ is a completely ramified value of $s(z)$. Also $r + n_1 \deg h = q, 0 < r < q$. It follows that every root of $s = 0$ has a multiplicity ≥ 2. Thus if $\deg h > 0$ and h has some zero with a multiplicity not divisible by q, then s has at least three completely ramified values. This is impossible. Hence we see that there are only two possible cases that arise

(i) $s^r h(s^{n_1}) = s^q$ or

(ii) $q|\mu(a)$ for every zero of h . \hfill (3.41)

However, case (i) implies $s^r h(s^{n_1}) = s^r (H(s^{n_1}))^q$ where H is a polynomial. It also implies that $q = r + n_1 \deg h = r + n_1 q \deg H$. This is impossible for $\deg H > 0$. Thus only case (ii), i.e., Eq. (3.41) can hold. Therefore, Eq. (3.37) becomes

$$F_0(z) = s(z)^{n_1 q} .$$

Letting $k = 2$ in Eq. (3.40), we get

$$[s_2^{r_2} h_2(s_2^q)]^q = s^{n_1 q} ,$$

and we may also assume

$$s_2^{r_2} h(s_2^q) = s^{n_1}; \quad r_2 + q \deg h_2 = n_2 .$$

Substituting s by s_2 and G by s, and exchanging the positions of n_1 and q, we can, using the reasoning above, obtain

$$h_2(s_2^q) = (K_2(s_2^q))^{n_1} , \quad r_2 + n_1 q \deg K_2 = n_2 ,$$

where K_2 is a polynomial. Since $r_2 > 0$ and $n_2 \le n_1 q$, $(n_2, q) = 1$, we conclude

$$s_2^{r_2} h_2(s_2^q) = s_2^{n_2} .$$

Hence,

$$F_0(z) = (s_3^{r_3} h_3(s_3^q))^q = s_2^{n_2 q} .$$

By repeating this kind of argument, we are led to

$$F_0(z) = s_2^{n_2 q} = s_3^{n_3 q} = s_4^{n_4 q} = \ldots .$$

This shows it is impossible for $F_0(z)$ to possess any zeros. Thus

$$F_0(z) = \lambda_0 F(z) = e^{H(z)}; \quad H(z) \text{ an entire function } .$$

Since we have shown that F has form (3.23), Theorem 3.14 is proven. This also concludes the proof of Theorem 3.12.

3.5. FACTORIZATION OF ELLIPTIC FUNCTIONS

In the previous section 3.4 we discussed the possible factors of the periodic cosine function and realized that they are quite restricted. Now we shall study the possible factors of an elliptic function $h(z)$. More specifically, we would like to find when $h(z) = f(g(z))$ what forms and properties, f (left factor) and g (right factor) may possess.

Theorem 3.15. No elliptic function $h(z)$ may have a periodic left factor.

Proof. First we recall that the order and lower order of an elliptic function h are both equal to 2. This is because any elliptic function can be expressed as a rational function of a sigma function and its derivatives.

Now the sigma function (and hence its derivatives) has order 2, so that the order of h $(= \rho(h)) \leq 2$. On the other hand, to any value "a",

$$T\left(r, \frac{1}{h-a}\right) \geq N\left(r, \frac{1}{h-a}\right) > \frac{cr^2}{\log r}$$

for some constant c. This leads to the conclusion that $\rho(h) \geq 2$. Therefore we arrive at $\rho(h) = 2$. We now suppose that h is not prime and has the factorization $h = f \circ g$. If f and g are both transcendental entire functions then, by Pólya's theorem, $\rho(f) = 0$, and f cannot be a periodic function. Also if f is periodic, say period 1, then g must be a polynomial. Consider a point set $S = \{z | z$ is a root of one of the equations $g(z) = m + c, m = 0, \pm 1, \pm 2, \ldots ; c$ is a constant$\}$. We shall show that S has one finite limit point. Let $g(z) = A_k z^k + \ldots + A_1 z + A_0$ and z_m, z_{m+j} be the roots satisfying $g(z) = m + c, g(z) = m + j + c$ respectively. We have

$$
\begin{aligned}
|g(z_m) - g(z_{m+j})| = |z_{m+j} - z_m||A_k||z_{m+j}^{k-1} + z_m z_{m+j}^{k-2} + \ldots \\
+ z_m^{k-1} + P_{k-2}(z_m, z_{m+j})| = j ,
\end{aligned}
\tag{3.42}
$$

where P_{k-2} is a polynomial with variables z_m and z_{m+j} of degree $k-2$ at most. We can easily see that for each m, there always exists integers $j_1, j_2 \geq m$ such that $|j_1 - j_2| < 4k^2$ and $|\arg z_1 - \arg z_2| \leq \frac{2\pi}{4k^2}$. We can thus derive

$$|z_{m+j}^{k-1} + \ldots + P_{k-2}(z_m, z_{m+j})| > \max(|z_{m+j}|^{k-1}, |z_m|^{k-1}) - |P_{k-2}|$$

which approaches ∞ as $m \to \infty$. This would imply from Eq. (3.42) that $|z_{m+j} - z_m|$ tends to 0 as $m \to \infty$. Therefore, S must have a limit point. Now we have $h = f \circ P$ where P is a polynomial. Assume that h has periods τ_1 and τ_2 and recall that f has period 1, then the following identity holds for any integers m, n_1 and n_2.

$$f(g(z + n_1\tau_1 + n_2\tau_2) + m) = f(g(z)) .
\tag{3.43}$$

For any fixed z_0 we set

$$S = \{z | g(z + n_1\tau_1 + n_2\tau_2) + m = g(z_0), n_1, n_2 \text{ and } m \text{ are integers}\} .$$

From the above analysis, we see that S has a finite limit point. Thus Eq. (3.43) will yield a conclusion that f is a constant, which is a contradiction. The theorem is thus proven.

With regard to the right factors of elliptic functions we have the following result.

Theorem 3.16. Let $h(z)$ be an elliptic function of valence 2. If $h = f \circ g$ for some transcendental entire function f and an entire function g that is not a linear polynomial, then the right factor g must be either (i) a polynomial of degree 2 or (ii) of the form $A\cos(z + \tau) + B$, where A, B, and τ are constants.

Proof. By assuming that h is an elliptic function of valence 2, it means that h satisfies the following differential equation:

$$(h')^2 = P(h) ,$$

where P is a polynomial. Thus we have

$$(g' f'(g))^2 = P(f(g)) .$$

If follows that

$$g'^2 = \frac{P(f(g))}{(f'(g))^2} = F(g) ,$$

where $F(\varsigma) = P(f(\varsigma))/f'(\varsigma)^2$, a meromorphic function.

According to a theorem of Clunie's, we conclude that $F(\varsigma)$ must be a rational function. Otherwise $\displaystyle\lim_{r\to\infty} \frac{T(r, F(g))}{T(r, g)} = \infty$ will result in $\displaystyle\varlimsup_{r\to\infty} \frac{T(r, F(g))}{T(r, g'^2)} = \infty$.

Therefore, we have

$$g'^2 = c \frac{(g - a_1)^{n_1} \ldots (g - a_k)^{n_k}}{(g - b_1)^{m_1} \ldots (g - b_l)^{m_l}} , \qquad c \text{ is a constant} \neq 0 . \tag{3.44}$$

Note that since g' has no poles, the denominator in the above expression never vanishes. Moreover, by examining the multiplicities of the roots of $g(z) = a_i$, we can verify easily that g has at most two complete ramified values, say a_1 and a_2. Then Eq. (3.44) becomes

$$g'^2 = c(g - a_1)^{n_1} (g - a_2)^{n_2} .$$

From this we conclude that if g is a polynomial then deg $g = 2$, and $n_1 = n_2 = 1$ if g is a transcendental entire function. In the latter we have

$g'^2 = d_1(g(z) + d_2)^2 + d_3$, where d_1, d_2, and d_3 are constants. It follows that $g(z)$ has the form $A\cos(cz + \tau) + B$. This also completes the proof.

Remark. It is not difficult to exhibit some elliptic functions have transcendental right factors. $S_n(z), C_n(z)$ and $d_n(z)$ are such functions.

$$S_n(2kz/\pi) = c\sin z \prod_{n=1}^{\infty} \left(\frac{1 - 2q^{2n}\cos 2z + q^{4n}}{1 - 2q^{2n-1}\cos 2z + q^{4n-2}} \right) .$$

We easily see that $S_n(2kz/\pi) = f(\sin z)$, where

$$f(\varsigma) = c\varsigma \prod_{n=1}^{\infty} \left(\frac{1 - 2q^{2n}(1 - 2\varsigma^2) + q^{4n}}{1 - 2q^{2n-1}(1 - 2\varsigma^2) + q^{4n-2}} \right) , \text{ a meromorphic function .}$$

Earlier in this chapter, we showed that if f is a transcendental entire function and P an arbitrary polynomial of degree ≥ 3, then $f \circ P$ cannot be periodic. We conclude this section by proving the following result.

Theorem 3.17. Let $f(z)$ be a non-constant meromorphic function and $P(z)$ be a polynomial of degree n. Then $F(z) = f \circ P(z)$ cannot be periodic unless $n = 1, 2, 3, 4$ and 6.

Proof. It is, of course, possible for $n = 1$ and 2. Therefore we shall only deal with the cases when $n \geq 3$.

Suppose that F is a periodic function and by changing variables if necessary, we may assume without loss of generality that F has a period of 1. Moreover, we may assume $P(z)$ has the form

$$P(z) = a_0 z^n + a_{n-t} z^{n-t} + a_{n-t-1} z^{n-t-1} + \ldots + \ldots$$

where t is an integer ≥ 2. It is clear that for any given z and the following equation

$$P(\varsigma) = P(z + m); \quad |z| > r_0 , \tag{3.45}$$

always has a root. Furthermore, for sufficiently large m, we have

$$\varsigma = \eta(z + m) + o(1) \quad (m \to \infty) , \tag{3.46}$$

where $\eta = e^{2\pi i/n}$. We observe that, for sufficiently large m, any integer m' (from the above we have $|\varsigma + m'|$) will be greater than r_0 as in Eq. (3.45). Since F has period 1,

$$F(\varsigma) = f(P(\varsigma)) = f(P(z + m)) = f(P(z)) = F(z) .$$

On the other hand, to the ς in Eq. (3.46), the equation

$$P(\varsigma') = P(\varsigma + m')$$

has a root ς' satisfying $\varsigma' = \eta(\varsigma + m') + o(1) = \eta^2 z + \eta^2 m + \eta m' + o(1)$ as $m \to \infty$. Thus

$$F(\varsigma' + m) = F(\varsigma') = F(\varsigma) = F(z) \ .$$

Consequently to a given point z, the equation

$$F(w) = F(z) \tag{3.47}$$

always has a root w satisfying

$$w = \eta^2 z + \eta(\eta m + m' + \eta^{-1}m) + o(1)$$

for any given integer m' and $|m|$ sufficiently large.

Since $\eta = e^{2\pi i/n}$, we have

$$\eta m + \eta^{-1}m + m' = \left(2\cos\frac{2\pi}{n}\right)m + m' \ .$$

Suppose that $2\cos\frac{2\pi}{n}$ is a irrational number. To any given real number β, we can choose m and m' such that the right-hand side of the above expression can be made arbitrarily close to β. Thus

$$F(\eta^2 z + \eta\beta) = F(z) \quad (-\infty < \beta < \infty) \ .$$

However, z is an arbitrarily given number, hence, from the above equation, F must be a constant function. This creates a contradiction. Suppose that $\cos\frac{2\pi}{n} = \alpha$ is a rational number. Then the nth root of unity η will satisfy the equation

$$\eta^2 - 2\alpha\eta + 1 = 0 \ .$$

In the meantime, η must satisfy an irredicuble equation $g(\eta) = 0$, where the degree of g is $\varphi(n)$ ($\varphi(n)$ denotes the Euler function of n). Therefore, $\eta^2 - 2\alpha\eta + 1$ must be divisible by $g(\eta)$. Hence $\varphi(n) = 1$ or 2. However,

$$\varphi(n) = n\prod_{q|n}\left(1 - \frac{1}{q}\right) \geq \prod_{q|n}(q - 1) \ .$$

It follows that if $\varphi(n) \leq 2$, then n can only have 2 and 3 as its factors. This results in $n = 3, 4$ *or* 6. The theorem is thus proved.

Discussion. Illustrate by examples, that for $n = 3, 4$ or 6 there exists a meromorphic function f_n and polynomial $P_n(z)$ of degree n such that

$$f_n(P_n(z)) \text{ is an elliptic function.}$$

3.6. FUNCTIONAL EQUATIONS OF CERTAIN MEROMOPRHIC FUNCTIONS

Factorization theory can be included in the theory of functional equations. The factorization of $F(z) = f(g(z))$ can be viewed as the finding of functions F, f and g that will satisfy the expression just mentioned. Various forms of functional equations have been derived in the course of studying problems relating physics, practical or theoretical mathematics. For example, a well-known problem is Cauchy's functional equation: $f(x + y) = f(x) + f(y)$. In general, it is difficult to obtain a concrete solution to a functional equation. Many have obtained results and focused their research on the necessary and sufficient conditions for the existence of solutions or certain special properties of the solutions. Here we shall introduce certain simple forms of the functional equations of meromorphic functions to show the existence as well as the growth properties of the solutions.

We shall first discuss the following type of equation:

$$f(g(z)) = h(z)f(z) , \qquad (3.48)$$

where we restrict f, g and h to entire functions. It is easy to derive the following results:

Theorem 3.18. Let $f(z), g(z)$ and $h(z)$ be entire functions and satisfy the Eq. (3.48). Suppose that $h(z)$ is a polynomial and g is not a linear polynomial. Then f must be a polynomial.

Theorem 3.19. Let $f(z), g(z)$ and $h(z)$ be entire functions and satisfy Eq. (3.48). Suppose that both f and h are non-constant polynomials, then g must be be a polynomial.

Theorem 3.20. Let $g(z) \equiv z^2$ and f, g be non-constant entire functions. Assume that h has only a finite number of zeros. If Eq. (3.48) holds for such f, g, and h, then f must have only a finite number of zeros as well.

Most of the results introduced here were obtained by R. Goldstein who also extended the previous discussion by considering meromorphic solutions of the following type of equation:

$$f(g(z)) = h(z)f(z) + H(z) . \qquad (3.49)$$

Theorem 3.21. Let f, g, h and H be meromorphic functions satisfying Eq. (3.49). Suppose that f, g are non-constant functions and g is always a transcendental entire function unless f is a rational function. Also suppose that there exists a positive constant k such that for $r > r_0$ (a constant),

$$\begin{aligned} T(r,h) &\le kT(r,f) , \\ T(r,H) &\le kT(r,f) . \end{aligned} \qquad (3.50)$$

Then g must be a rational function of, say, order m. Furthermore, if f is transcendental, then $1 \le m \le k+1$, and when $m > 1$, f must satisfy

$$T(r,f) = O((\log r)^{\alpha+\epsilon}) ,$$

where ϵ is any given positive number and $\alpha = \log(2k+1)/\log m$.

Proof. From Eqs. (3.49) and (3.50), we have

$$\begin{aligned} T(r,f(g)) &\le T(r,f) + T(r,h) + T(r,H) + O(1) \\ &< (2k+1)T(r,f) + O(1) . \end{aligned} \qquad (3.51)$$

But Clunie proved that if f and g are transcendental, then

$$\varlimsup_{r\to\infty} \frac{T(r,f(g))}{T(r,f)} = \infty \qquad (3.52)$$

which will contradict with Eq. (3.51). Hence f and g cannot both be transcendental. It is clear that if f is rational function and g is transcendental, then from Eq. (3.50) we conclude both h and H are rational functions. As a result Eq. (3.49) will not hold. For this equation to hold, f must be transcendental and g must be a polynomial.

Before proceeding further we prove the following result.

Lemma 3.8. Let $\psi(r)$ be a positive and continuous function of r satisfying, for some $m > 1$,

$$\psi(\mu r^m) \le A\psi(r) \quad (r \ge r_0) , \qquad (3.53)$$

where $\mu, A(A > 1)$ are two positive constants. Then

$$\psi(r) = O((\log r)^\alpha); \quad \alpha = \log A / \log m .$$

Proof. We put in Eq. (3.53)

$$t = \log r - \frac{\log \mu}{1 - m}, \quad \psi(r) = \phi(t)$$

that yields

$$\phi(mt) \le A\phi(t), \quad (t \ge t_0) .$$

We choose α such that $m^\alpha = A$ and put $\psi(t) = \phi(t)/t^\alpha$. Then Eq. (3.53) becomes

$$\varphi(mt) \le \varphi(t), \quad (t \ge t_0) .$$

Now, $\varphi(t)$ is also a positive and continuous function for $t > 0$ and the above inequality ensures that it is bounded above by some number B for sufficiently large values of t. Thus

$$\psi(r) = \phi(t) \le Bt^\alpha = B \left(\log r - \frac{\log \mu}{1 - m} \right)^\alpha$$

$$\le B_1 (\log r)^\alpha; \quad \text{where } B_1 \text{ is a suitable constant.}$$

Now we continue the proof of Theorem 3.21.

Let

$$g(z) = a_m z^m + a_{m-1} z^{m-1} + \ldots + a_0 \quad (a_m \ne 0), \quad m \ge 2 ,$$

and $\mu = |a_m| - \delta \; (0 < \delta < |a_m|)$. Since $|g(z)| \sim |a_m| r^m$ for sufficiently large values r, we have for any value "a".

$$n(r, a, f(g)) \ge mn(\mu r^m, a, f), \quad (r \ge r_0) .$$

By integrating, we get

$$N(r, a, f(g)) - N(r_0, a, f(g)) \ge \int_{r_0}^r \frac{mn(\mu t^m, a, f)}{r} dt + O(1) \log r .$$

We put $s = \mu t^m$ and obtain

$$\int_{r_0}^r \frac{mn(\mu t^m, a, f)}{t} dt = \int_{\mu r_0^m}^{\mu r^m} \frac{n(s, a, f)}{s} ds$$

$$= N(\mu r^m, a, f) - N(\mu r_0^m, a, f) + O(\log r) .$$

By combining the above two inequalities, we have

$$N(\mu r^m, a, f) + O(\log r) \leq N(r, a, f(g)), \quad (r \geq n_0) . \tag{3.54}$$

But, it is well known that for a suitable value a,

$$N(r, a, f) \sim T(r, f)$$

and $\log r = o(T(r, f))$ for a transcendental function f. By Eqs. (3.54) and (3.51) we obtain

$$T(\mu r^m, f) \leq (2k + 1 + \varepsilon) T(r, f), \quad r \geq r_1 .$$

Applying Lemma 3.8, the required result follows.

By a similar argument we can obtain the following result:

Theorem 3.22. Let f, g and h be non-constant meromorphic functions satisfying Eq. (3.48). Suppose that g is always a transcendental function unless f is rational. If there exists a positive constant k such that

$$T(r, h) \leq kT(r, f), \quad (r \geq r_0)$$

then g must be a rational function of order m. Furthermore, if $m > 1$ and ε is any given positive number, then unless f is rational, $m \leq k + 1$, and

$$T(r, f) = O(\log r)^{\beta + \varepsilon} \quad \text{as } r \to \infty ,$$

where $\beta = \log(k + 1) / \log m$, and

$$\frac{T(r, h)}{T(r, f)} > m - 1 - \varepsilon \quad (r \geq r_1) . \tag{3.55}$$

In the following we shall investigate Eq. (3.48), in which the zeros or poles of h have been restricted.

We shall call the value a a Fatou exceptional value of g of multiplicity m if $g(z) \equiv a + (z - a)^m e^{G(z)}$; where $G(z)$ is an entire function.

Theorem 3.23. Let f and h be meromorphic functions, and g be a nonlinear entire function satisfying Eq. (3.48). Suppose that h has no poles (zeros), then f has at most one pole (zero) at $z = \alpha$, and a is a Fatou exceptional value of g of multiplicity 1.

Proof. We give the proof for the case where $h(z)$ has no poles. The case where h has no zeros can be proved similarly. Using the assumption, we easily see that if $z = a$ is a pole of $f(z)$ and $g(b) = a$, then $z = b$ must be a pole of $f(z)$. Repeating this argument yields if $z = a$ is a pole of f and if for some n, $g_n(z)$ (the nth iterate of g) $= a$, then $z = b$ is also a pole of $f(z)$. We now need a result of Fatou's as follows.

Lemma 3.9. Let $g(z)$ be a nonlinear entire function. Then there exists a nonempty perfect set $T(= T(g))$ of complex numbers with the property that to any $z_0 \in T$ and an arbitrary number w (with one possible exceptional value) there corresponds a sequence of positive integers $\{n_j\}(j = 1, 2, \ldots)$ and a sequence of complex numbers $\{z_j\}(j = 1, 2, \ldots)$ such that

$$\lim z_j = z_0$$

and

$$g_{n_j}(z_j) = w, \quad (j = 1, 2 \ldots) . \tag{3.56}$$

This result combined with the conclusion at the beginning of the proof, implies that f has at most one pole, at $z = a$. Furthermore, $z = a$ must be a Fatou exceptional value of $g(z)$. Hence we have

$$g(z) = a + (z - a)^m e^{G(z)} , \tag{3.57}$$

where $G(z)$ is an entire function and m is a non-negative integer to be determined. We express $f(z)$ as

$$f(z) = \frac{F(z - a)}{(z - a)^n} , \tag{3.58}$$

where F is an entire function with $F(0) \neq 0$ and n is a positive integer. We will only treat the case where $m \geq 1$ (a similar argument applies if $m = 0$). Then Eqs. (3.48), (3.57), and (3.58) yields

$$\frac{F\{(z - a)^m e^{G(z)}\}}{(z - a)^{nm} e^{nG(z)}} = h(z) \frac{F(z - a)}{(z - a)^n} .$$

Consideration of the order of the pole at $z = a$ leads to either $n = 0$ or $m = 1$. But when $n = 0$, f becomes an entire function which contradicts the hypothesis that $z = a$ is a pole of f. Therefore $m = 1$ and f and g are given by Eqs. (3.58), (3.57) respectively. Theorem 3.23 is thus proven.

Corollary 3.1. Let f, g and h be as in Theorem 3.23. If $g(z)$ and $h(z)$ are nonlinear polynomials, then $f(z)$ has no poles (zeros).

Theorem 3.24. Let $g(z)$ be a polynomial of degree $m \geq 2$, and let f, h be meromorphic functions satisfying the equation $f(g(z)) = h(z)f(z)$. Suppose that f is of finite order and $\delta(0, h) = \delta(\infty, h) = 1$. Then $\delta(0, f) = \delta(\infty, f) = 1$ and $\rho_h = m\rho_f$; where ρ_h and ρ_f are the orders of f, h respectively and must be positive integers.

Proof. It is well known under the hypotheses that

$$\rho_{f(g)} = m\rho_f . \tag{3.59}$$

By $\delta(0, h) = \delta(\infty, h) = 1$, h is of regular growth and is of positive integer or infinite order. Now

$$h(z) = \frac{f(g(z))}{f(z)} ,$$

hence,

$$1 \leq \rho_h \leq \max(\rho_{f-1}, \rho_{f(g)}) = \max(\rho_f, \rho_{f(g)}) = \max(\rho_f, m\rho_f) .$$

Thus

$$\rho_f > 0 \quad \text{and}$$
$$\rho_h \leq m\rho_f < \infty . \tag{3.60}$$

On the other hand, from the equation $f(g) = hf$ and Eq. (3.59) we have

$$m\rho_f = \rho_{f(g)} \leq \max(\rho_h, \rho_f) . \tag{3.61}$$

Since $\rho_f < \infty$, we deduce from the above inequality that

$$m\rho_f \leq \rho_h . \tag{3.62}$$

Thus, by Eqs. (3.60) and (3.62),

$$\rho_h = m\rho_f = \rho_{f(g)} < \infty . \tag{3.63}$$

From this and the fact that h is of regular growth we have

$$T(r, f) < r^{\rho_f + \epsilon} < r^{\rho_h - \epsilon} < T(r, h), \quad (r \geq r_0)$$

whence

$$\lim_{r\to\infty} \frac{T(r,h)}{T(r,f)} = \infty \ . \tag{3.64}$$

From the equation $f(g) = hf$, we have

$$T(r, f \circ g) \le T(r,h) + T(r,f) \ .$$

Hence from Eq. (3.64) we get

$$\frac{T(r, f \circ g)}{T(r,h)} \le 1 + o(1) \quad (\text{as } r \to \infty) \ .$$

Conversely, from $h = f(g)/f$ and Eq. (3.64) we deduce

$$\frac{T(r, f(g))}{T(r,h)} \ge 1 + o(1) \ .$$

Thus, we have

$$T(r, f(g)) \sim T(r,h) \ . \tag{3.65}$$

Now $n(r, 0, f(g)) \le n(r, 0, h) + n(r, 0, f)$, so

$$\begin{aligned} N(r, 0, f(g)) &\le N(r, 0, h) + N(r, 0, f) + O(\log r) \\ &\le N(r, 0, h) + T(r, f) + O(\log r) \ . \end{aligned}$$

Using Eq. (3.64) and the assumption $\delta(0, h) = 1$, we have

$$\overline{\lim} \frac{N(r, 0, f(g))}{T(r,h)} \le \overline{\lim_{r\to\infty}} \frac{N(r, 0, h)}{T(r,h)} + o(1) = o(1) \ . \tag{3.66}$$

Therefore, by Eqs. (3.66) and (3.65) we have

$$\overline{\lim_{r\to\infty}} \frac{N(r, 0, f(g))}{T(r, f(g))} = \overline{\lim_{r\to\infty}} \frac{N(r, 0, f(g))}{T(r, h)} = o(1) \ .$$

This shows that $\delta(0, f(g)) = 1$. Similarly we can prove that $\delta(\infty, f(g)) = 1$. Now we prove that ρ_f must be a positive integer. If $f \circ g$ is a meromorphic function of finite order, and g is a nonlinear polynomial satisfying $f(g) = hf$, then by $\delta(0, f(g)) = \delta(\infty, f(g)) = 1$, we also have $\delta(0, f) = \delta(\infty, f) = 1$. (For the proof, we refer the reader to Goldstein's paper "Some results on factorization of meromorphic functions", *J. London Math. Soc.* (2) **4** (1971) 357-364.

Hence, ρ_f must be a positive integer. This also completes the proof of the theorem.

Discussion. (i) If $g(z)$ is allowed to be meromorphic and h has no poles, what conclusions will result? (ii) Does the condition $\delta(0, f(g)) = \delta(\infty, f(g)) = 1$ always lead to $\delta(0, f) = \delta(\infty, f) = 1$?

Theorem 3.25. Let f be a non-constant meromorphic function, g be an entire function, and $q(z)$ be a polynomial of degree k (≥ 1) satisfying the following equation

$$f(g) = q(f) .$$

Then $q(z)$ must be a polynomial of degree $m \leq k$. Furthermore, if $m > 1$, then $T(r, f) = O(1)(\log r)^\alpha$, where

$$\alpha = (\log k / \log m) + \varepsilon \quad (\varepsilon \text{ is any given positive number}) .$$

We omit the proof, this being analogous to the proof of Theorem 3.21.

3.7. UNIQUENESS OF FACTORIZATION

For simplicity we shall only discuss entire functions and their entire factors. We state that two factorizations (of entire factors) $f_1(f_2(\ldots (f_n))\ldots)$ and $g_1(g_2(\ldots (g_n))\ldots)$ are equivalent if there exists linear transformations $\lambda_1, \ldots, \lambda_{n-1}$ such that

$$f_1 = g_1(\lambda_1), f_2 = \lambda_1^{-1}(g_2(\lambda_2)), \ldots, f_n = \lambda_{n-1}^{-1}(g_n) .$$

An entire function F is called uniquely factorizable, if all its factorizations of nonlinear prime entire factors are equivalent to each other. Ritt obtained a complete answer to the uniqueness factorization problem for polynomials (see the appendix). The result essentially states that, besides the following three non-equivalent cases, for pairs of consecutive factors $f_1(f_2)$ and $g_1(g_2)$, the two factorizations of a polynomial $F(z)$ will be equivalent. The exceptions are:

(i) $f_1(z) = z^k$, $f_2(z) = z^l$ and $g_1(z) = z^l$, $g_2(z) = z^k$

(ii) $f_1(z) = z^k[h(z)]^l$, $f_2(z) = z^l$, and $g_1(z) = z^l$, $g_2(z) = z^k h(z^l)$,

and

(iii) $f_1(z) = P_k(z)$, $f_2(k) = P_l(z)$, and $g_1(z) = P_l(z)$, $g_2(z) = P_k(z)$,

where $P_k(z)$ is the kth polynomial satisfying $\cos kz = P_k(\cos z)$.

Case (ii) may also arise in the factorization of a transcendental entire function. If, for example, we let $F(z)$ be $z^p \exp z^p$ (p is a prime number), then F has two factorizations that are not equivalent: $F(z) = z^p \circ (ze^{z^p/p})$ and $F(z) = (ze^z) \circ z^p$. However, it is not difficult to show that both $F(z) = z^p e^{pz} (= z^p \circ ze^z)$ and $F(z) = (ze^z) \circ (ze^z)$ are uniquely factorizable. Moreover, the latter is almost the simplest function one can show in demonstrating the uniqueness factorization of transcendental entire factors.

As a generalization, H. Urabe obtained the following result in his dissertation.

Theorem 3.26. Let $F(z) = (ze^z) \circ (h(z)e^z)$, where $h(z)$ is a nonconstant entire function of order less than one, and has at least one simple zero. Then F is uniquely factorizable.

Proof. (sketch) Let $F(z) = (ze^z) \circ (h(z)e^z) = f(g(z))$; f, g being two nonlinear entire functions. By virtue of the assumption that $h(z)$ has at least one simple zero and the Tumura-Clunie Theorem we conclude that f cannot be a polynomial. According to a result of Edrei-Fuchs' that if f and g are two transcendental entire functions with the exponent of convergence of the zeros of f being positive, then the zeros of $f(g)$ have an exponent of convergence equal to infinity. Therefore, we need only consider three cases: (a) $f(z) = h_1(z)e^{p(z)}$, h_1 nonlinear entire function with $\rho(h) = 0$, $p(z)$ is a non-constant polynomial, and $g(z)$ is a transcendental entire function with $\rho(g) < 1$; (b) $f(z) = ze^{p(z)}$, $g(z) = h(z)e^{q(z)}$, where p, and q are non-constant entire functions; and (c) $f(z) = h_1(z)e^{p(z)}$, where $g(z)$ is a polynomial of degree ≥ 2, and h_1 and p are non-constant entire functions satisfying $\rho(h_1) < \frac{1}{\deg g}$ (hence $\rho(h_1)(g) < 1$).

In case (a), from $F = f(g)$, we obtain $h_1(g(z)) = h(z)e^{d(z)}$ and $p(g(z)) = z - d(z) + h(z)e^z$, where $d(z)$ is an entire function with $\rho(d) < 1$ (Polya's Theorem, Corollary A.1, Appendix). By Theorem 4.2, it follows that $p(z)$ must be a polynomial. Thus $\rho(g) = 1$, which is a contradiction. In case (b), we obtain a functional equation $q(z) + p(h(z)e^{q(z)}) = z + h(z)e^z$. It is easily verified from this relationship that $q(z)$ must be linear and the uniqueness of factorization follows. In case (c), we have the relations $h_1(g(z)) = h(z)$ and $p(g(z)) = z + h(z)e^z$. Using these equations and an argument similar to the proof of case (b) we easily arrive at the uniqueness of factorization of F.

Urabe also obtained the following more general result.

Theorem 3.27. Let $F(z) = (z + h(e^z)) \circ (z + q(e^z))$, where $h(z)$ is a non-constant entire function with the order $\rho(h(e^z)) < \infty$ and $q(z)$ is a non-constant polynomial. Then $F(z)$ is uniquely factorizable.

We note that e^z and $\cos z$ are pseudo-prime and have an infinite number of different factorizations.

There exist some transcendental entire functions that are not pseudo-prime but have an infinite number of equivalent factorizations. For example, if we let $F(z) = z - \sin z + \sin(\sin z - z)$. Then $F(z) = f \circ f = g \circ g$ where $f(z) = z - \sin z$ and $g(z) = \sin z - z + 2k\pi$ (k integer $\neq 0$).

The following questions are therefore interesting.

Question 1. Do there exist two nonequivalent factorizations $f_1 \circ f_2 = g_1 \circ g_2$, where f_1, f_2, g_1, g_2 are prime transcendental entire functions?

Question 2. (Gross) Do there exist prime nonlinear entire functions $f_1, f_2, \ldots, f_m$ and $g_1, \ldots, g_n$ with $n \neq m$ such that

$$f_1 \circ f_2 \circ \ldots \circ f_m \equiv g_1 \circ g_2 \circ \ldots \circ g_n?$$

4

FIX-POINTS AND THEORY OF FACTORIZATION

4.1. THE RELATIONSHIP BETWEEN THE FIX-POINTS AND THEORY OF FACTORIZATION

We have shown in the previous chapter that $e^z + z$ is prime. Gross conjectured that functions $F(z)$ of the form

$$F(z) = Q(z)e^{\alpha(z)} + z , \qquad (4.1)$$

where $Q(z)$ is a polynomial and $\alpha(z)$ is a non-constant entire function, must be prime.

To date, the conjecture has not been answered.[*] However, some partial results have been obtained. Most of these were stated in terms of fix-points. Recall that at the beginning of Chapter 3 we proved that if $P(z)$ is a nonlinear polynomial and f is a transcendental entire function, then $P(f(z))$ has an infinite number of fix-points (Theorem 3.2).

We now prove the following lemma.

Lemma 4.1. Let f and g be two non-constant entire functions. If $f(g)$ has only a finite number of fix-points then $g(f)$ also has only a finite number of fix-points.

Proof. Let z_0 be a fix-point of $f(g)$. That is, if $f(g(z_0)) = z_0$, then $g(f(g(z_0))) = g(z_0)$. Thus $g(z_0)$ is a fix-point of $g(f)$. Moreover, if z_1 and z_2

---

[*]Recently (1988) W. Bergweiler confirmed this (and hence conjecture 1 in next page) in his preprint entitled "Proof of a conjecture of Gross concerning fix-points" by utilizing Wiman-Valiron type of argument.

149

are two distinct fix-points of $f(g)$, then $g(z_1)$ and $g(z_2)$ will be two distinct fix-points of $g(f)$. If $g(z_1) = g(z_2)$, then $z_1 = f(g(z_1)) = f(g(z_2)) = z_2$. This proves the lemma.

From this lemma and Theorem 3.2 we conclude that $f(P(z))$ has an infinite number of fix-points for any transcendental entire function f and nonlinear polynomial $P(z)$.

We also know that if f, g are two transcendental entire functions, then either g or $f(g)$ must have infinite number of fix-points. It is easy to see then that Gross' conjecture is equivalent to the following conjecture:

Conjecture 1. If f and g are two nonlinear entire functions, with at least one of them being transcendental, then $f(g)$ must have an infinite number of fix-points.

4.2. CONJECTURE 1 WITH $\rho(f(g)) < \infty$

Conjecture 1 with the additional hypothesis that $\rho(f(g)) < \infty$, has been validated by Goldstein, Yang and Gross, and Prokopovich, each using different approach and frames. We now exhibit Prokopovich's statement and its proof as follow:

Theorem 4.1. Let

$$F(z) = Q_1(z) + Q_2(z)e^{P(z)} \qquad (4.2)$$

where Q_1, Q_2 and P are polynomials with $Q_1(z) \not\equiv$ constant. $P(z) \not\equiv$ constant, and $Q_2(z) \not\equiv 0$. Then $F(z)$ is prime unless there exist polynomials q_1, q_2, q_3, T, U, V and nonlinear polynomial $g(z)$ such that $F = f(g)$ with

$$Q_1(z) = q_1(g(z)), \quad Q_2(z) = q_2(g(z)), \quad p(z) = q_3(g(z)) \, ,$$
and $\quad f(z) = T(z) + U(z)e^{V(z)} \, .$

Before proving the theorem, we first quote the following useful fact.

Lemma 4.2. Let $P(z)$ and $Q(z)$ be polynomials of degrees p, q respectively. Suppose that $q \geq 2, q|p$, and that $P(z)$ is not a function of $Q(z)$. Then $P(z)$ can be expressed as

$$P(z) = P_1(Q(z)) + P_2(z) \, , \qquad (4.3)$$

where P_1 and P_2 are polynomials and $q \nmid \deg P_2$.

Proof of Theorem 4.1. First of all, it is not difficult to show that F is not a periodic function. Hence it suffices to show that F is E-prime.

Suppose that

$$F(z) = f(g(z)) , \tag{4.4}$$

where f and g are nonlinear entire functions.

Differentiating Eq. 2.1, we obtain

$$F(z) = f'(g(z))g'(z) = Q_1'(z) + S(z)e^{P(z)}, \quad S(z) \equiv Q_2'(z) + Q_2(z)P'(z) . \tag{4.5}$$

We now see that the function $\varphi(z) = F'(z)/Q_1'(z)$ has only a finite number of poles and 1-points. Since φ is not a rational function, $\varphi(z)$ and, hence, $F'(z)$ both have an infinite number of zeros. Let $z_1, z_2, \ldots z_n, \ldots$, be all the distant zeros of $F'(z)$. Then

$$F'(z_n) = Q_1'(z_n) + S(z_n)e^{P(z_n)} = 0 , \quad n = 1, 2, \ldots .$$

Thus

$$e^{P(z_n)} = -\frac{Q_1'(z_n)}{S(z_n)} .$$

It follows that

$$
\begin{aligned}
F(z_n) &= Q_1(z_n) + Q_2(z_n)e^{P(z_n)} \\
&= [S(z_n)Q_1(z_n) - Q_2(z_n)Q_1'(z_n)]/S(z_n) .
\end{aligned}
$$

Also from Eq. (4.5), we see that $\deg S \geq \deg Q_2$, and hence $t = \deg(SQ_1 - Q_2Q_1') > \deg S$. This shows that in $\{z_n\}$ there exist no more than $2t$ distinct points $z_{n_j}, j = 1, 2, \ldots, 2t$ such that

$$F(z_{n_1}) = F(z_{n_2}) = \ldots = F(z_{n_{2t}}) . \tag{4.6}$$

First we treat the case that g is a transcendental entire function. By a theorem of Pólya we know that f is of zero order. Suppose there exists a point ς_0 that is a zero of f' but not a Picard exceptional value of g. This implies that there exists an infinite sequence of points $\{z_j^*\}$ such that $g(z_j^*) = \varsigma_0$. But then $F'(z_j^*) = f'(\varsigma_0)g'(z_j^*) = 0, j = 1, 2, \ldots$. This shows that $\{z_j^*\} \subset \{z_n\}$. Furthermore, we have $F(z_j^*) = f(g(z_j)) = f(\varsigma_0)$, $j = 1, 2, \ldots$, which contradicts with the result in Eq. (4.6). Therefore, we conclude that f' can have only one zero, say ς_0. Moreover, ς_0 must be a Picard exceptional value of g. Since f' is of zero order, it follows that f' must be a polynomial with the form

$$f'(\varsigma) = A(\varsigma - \varsigma_0)^m, \quad m \geq 1 ,$$

and

$$g(z) = \varsigma_0 + a(z)e^{b(z)} \ ,$$

where $a(z)$ and $b(z)$ are polynomials with $b(z) \not\equiv$ constant. Thus,

$$F'(z) = Aa^m(z)[a'(z) + a(z)b'(z)]e^{(m+1)b(z)} \ .$$

This implies that $F'(z)$ has only a finite number of zeros, a contradiction.

Now what left is the case where $g(z)$ is a polynomial of degree $n \geq 2$. Assume that deg $Q_1 = \alpha$, deg $P = \beta$ and note that now $f'(\varsigma)$ must have an infinite number of zeros $\varsigma_1, \varsigma_2, \ldots$. Let $z_j^{(k)}, k = 0, 1, 2, \ldots$ be the roots of the equation $g(z) = \varsigma_j$.

It is easy to verify that when $0 \leq k \leq n - 1$ and j sufficiently large, we have

$$z_j^{(k)} = (1 + o(1))r_j e^{2\pi ik/n}, \quad r_j \to \infty \ (j \to \infty) \ . \tag{4.7}$$

Let j be fixed and z_j, z'_j be two points in $\{z_j^{(k)}\}$ corresponding to $k = 0$ and $k = 1$ respectively. Then

$$F'(z_j) = F'(z'_j) = 0, \quad S(z_j)e^{P(z_i)} = -Q_1(z_j)$$

and

$$S(z'_j)e^{P(z'_j)} = -Q_1(z'_j) \ .$$

By $F(z_j) = f(\varsigma_j) = F(z'_j)$, we have

$$Q_1(z_j) + Q_2(z_j)e^{P(z_j)} = Q_1(z'_j) + Q_2(z'_j)e^{P(z'_j)} \ .$$

Thus

$$\begin{aligned}
Q_1(z_j) - Q_1(z'_j) &= Q_2(z'_j)e^{P(z'_j)} - Q_2(z_j)e^{P(z_i)} \\
&= \frac{Q_2(z_j)Q'_1(z_j)}{S(z_j)} - \frac{Q_2(z'_j)Q'_1(z'_j)}{S(z'_j)} \ .
\end{aligned} \tag{4.8}$$

Three cases arise that are to be treated separately.

Case (i). $n \nmid \alpha$.

From Eq. (4.7), we have, for sufficiently large j,

$$Q_1(z_j) - Q_1(z'_j) = a(1 + o(1)r_j^\alpha(e^{2\pi i\alpha/n} - 1) \ , \tag{4.9}$$

where a is the coefficient of the leading term in $Q_1(z)$. It is also easily verified that

$$\frac{Q_2(z_j)Q_1'(z_j)}{S(z_j)} - \frac{Q_2(z_j')Q_1'(z_j')}{S(z_j')} = O(1)r_j^{\alpha-\beta}, \ j \to \infty . \tag{4.10}$$

Combining Eqs. (4.7) and (4.10) yields:

$$a(1 + o(1))(e^{2\pi i\alpha/n} - 1)r_j^{\beta} = O(1) .$$

This is clearly impossible. Thus case (i) is ruled out.

Case (ii). $n|\alpha$ and $Q_1(z)$ cannot be expressed as a polynomial in g. Then by Lemma 4.2, we have

$$Q_1(z) = T(g(z)) + U(z) ,$$

where T and U are polynomials with $n \nmid \deg U$. We put

$$\begin{aligned} G(z) &= F(z) - T(g(z)) = f(g(z)) - T(g(z)) \\ &= h(g(z)) = U(z) + Q_2(z)e^{P(z)} , \end{aligned}$$

where $h = f - T$. By applying the argument used in Case (i) to $G(z) = h(g(z))$, we arrive at a similar contradiction. Thus Case (ii) is ruled out.

Case (iii). $n|\alpha$ and $Q_1(z) = T(g(z))$, where T is a polynomial. From $Q_1(z_j) = Q_1(z_j') = T(\varsigma_j)$ and Eq. (4.8) we have

$$\frac{Q_2(z_j)T'(\varsigma_j)g'(z_j)}{S(z_j)} = \frac{Q_2(z_j')T'(\varsigma_j)g'(z_j')}{S(z_j')} . \tag{4.11}$$

Note that $T'(\varsigma) \not\equiv 0$ and $T'(\varsigma)$ has only a finite number of zeros.

It follows from the above equation that there exist an infinite number of j's such that

$$\frac{Q_2(z_j)g'(z_j)}{S(z_j)} = \frac{Q_2(z_j')g'(z_j')}{S(z_j')} . \tag{4.12}$$

For the rest of the proof, we may assume that $j \geq j_0$ (i.e., j is sufficiently large).

It is easily verified that

$$\frac{Q_2(z_j^{(k)})g'(z_j^{(k)})}{S(z_j^{(k)})} = \lambda(1 + o(1))r_j^{n-\beta}e^{-2\pi i k\beta/n} \ (j \to \infty) , \tag{4.13}$$

where λ is a constant $\neq 0$. If $n \nmid \beta$, then from the above two equations, we have

$$\lambda(1 + o(1))r_j^{n-\beta}(e^{-2\pi\beta/n} - 1) = 0 ,$$

which is impossible. Finally, if $n | \beta$ but $P(z)$ is not a polynomial g, then $P(z)$ can be expressed as $P(z) = V(g(z)) + W(z)$ where V, W are polynomials with $n \nmid \deg W$. Then, as before, a similar contradiction will result.

Summing up the above discussions, we may conclude that if $F(z) = Q_1(z) + Q_2(z)e^{P(z)}$ can be expressed as $f(g(z))$, where g is a polynomial, then $Q_1(z) = T(g(z))$ and $P(z) = V(g(z))$; where T and V are polynomials. Moreover,

$$Q_2(z) = \frac{f(g(z)) - T(g(z))}{e^{V(g(z))}} = U(g(z)) ,$$

where $U(\varsigma) = [f(\varsigma) - T(\varsigma)]e^{-V(\varsigma)}$. Since Q_2 is a polynomial, it follows that U is also a polynomial. This completes the proof of the theorem.

Conjecture 1 with the additional hypothesis that $\rho(f(g)) < \infty$ is proven in terms of fix-points as follows.

Corollary 4.1. Let f and g be two nonlinear entire functions with at least one of them being transcendental. If $\rho(f(g)) < \infty$, then $f(g)$ must have an infinite number of fix-points. This is equivalent to saying that $Q(z)e^{P(z)} + z$ is prime for any polynomial $Q(z)(\not\equiv 0)$ and $P(z)(\not\equiv$ constant$)$.

4.3. SOME GENERALIZATIONS

Theorem 4.2. (Goldstein) Let $P_m(z)$ be a polynomial of degree $m(\geq 1)$. Let $\phi_m(z)(\not\equiv$ constant$)$ and $\psi_m(z)(\not\equiv$ constant$)$ be entire functions of order $< m$. Suppose that $F(z) = \phi_m(z) + \psi_m(z)\exp(P_m(z)) = f(g(z))$ for some nonlinear, entire functions f and g. Then $g(z)$ is a polynomial of degree $k \leq m$. Moreover, if ρ_f denotes the order of f, then $k\rho_f = m$.

Almost concurrently Prokopovich obtained the following result.

Theorem 4.3. Let $F(z) = \phi_1(z) + \phi_2(z)e^{P(z)}$, where $P(z)$ is a nonconstant polynomial and $\phi_1(z)(\not\equiv$ constant$)$, $\phi_2(z)(\not\equiv 0)$ are two entire functions satisfying

$$T(r, \phi_j) = o(1)T(r, F), (j = 1, 2), \quad \text{as} \quad r \to \infty . \tag{4.14}$$

If $F(z)$ can be factorized as $F = f(g)$, where f and g are entire functions with g being a nonlinear polynomial satisfying $\deg g < \deg P$, then $\phi_1(z)$ and $\phi_2(z)e^{P(z)}$ have g as their common right factor.

Remark. We note that the condition in Eq. (4.14) of the present theorem is weaker than that in Theorem 4.2, but here g is confined to be a polynomial of degree less than the degree of P.

In order to prove the theorem we need to introduce some concepts, notations, and, lemmas about algebroid functions.

Definition 4.1. Let $a_0(z), a_1(z), \ldots, a_k(z)$ be $k+1$ meromorphic functions (in the z-plane). A solution $U = U(z)$ of the equation

$$P(U, z) = a_0(z)U^k + a_1(z)U^{k-1} + \ldots + a_k(z) = 0$$

is called a k-valued algebroidal function.

Here we do not require that the polynomial $P(U, z)$ be irreducible over the field of meromorphic functions. When all the $a_j(z)(j = 0, 1, 2, \ldots, k)$ are polynomials, $U(z)$ becomes the algebraic function.

Let $\{c_\nu\}$ denote the countable set of the branch points of the algebroidal function $U(z)$. Then, it is possible to draw from each point c_ν, a ray, l_ν such that these rays will not pass any poles of $U(z)$ and intersect one another. Let $\Gamma_U = \cup_\nu l_\nu$ and $D_U = (|z| < \infty) \backslash \Gamma_U$. Then D_U is a simply connected domain. In D_U the function $U(z)$ can be decomposed into k single-valued branches, i.e., $U_1(z), U_2(z), \ldots U_k(z) - k$ analytic functions. If all the functions are distinct, and, moreover, given any two branches one could be obtained from the other by analytic continuation along a suitable path, then we shall call such an algebroidal function $U(z)$ an exactly k-valued function. Otherwise, we call $U(z)$ a degenerated k-valued function.

H. Selberg introduced the standard Nevanlinna quantities, which are analog to the proximate function, counting function, and characteristic function in the Nevanlinna theory of meromorphic functions, for algebroidal functions as follows.

$$m(r, U) = \frac{1}{2\pi k} \sum_{i=1}^{k} \int_0^{2\pi} \log^+ |U_i(re^{i\theta})| d\theta , \qquad (4.15)$$

$$N(r, U) = \frac{1}{k} \int_1^r \frac{n(t, U) - n(0, U)}{t} dt + \frac{n(0, U)}{k} \log r \qquad (4.16)$$

and

$$T(r, U) = m(r, U) + N(r, U)$$

where k is the member of branches of the algebroidal function $U(z)$.

Clearly the "order" of an algebroidal function $U(z)$ can be defined in terms of $T(r, U)$ as in the Nevanlinna theory.

Lemma 4.3. (Selberg) Let $U(z)$ be a s-valued algebroidal function of finite order and let $a_\nu(z)(\nu = 1, 2, \ldots, 2s + 1)$ be arbitrary $2s + 1$ distinct complex numbers. Then

$$T(r, U) = \sum_{\nu=1}^{2s+1} N\left(r, \frac{1}{U(z) - a_\nu}\right) + O\log r , \quad r \to \infty$$

Lemma 4.4. (Prokopovich) Let $f(z)$ be a transcendental meromorphic function of finite order with $\delta(c, f) = 1$ for some value c $(|c| \le \infty)$. Let $U(z)$ be an exactly k-valued algebroidal function satisfying $T(r, U) = o(1)T(r, f)$ as $r \to \infty$. Then

$$N\left(r, \frac{1}{f - U}\right) \ge \left(1 - \frac{1}{k} + o(1)\right)T(r, f), \quad r \to \infty .$$

We shall also need the following fact.

Lemma 4.5. (Prokopovich) Let $g(z)$ be a transcendental meromorphic function satisfying $N(r, g) = oT(r, g)$ as $r \to \infty$. If $R(z)$ is a rational function of order k, and $\alpha(z)$ is a non-constant meromorphic function satisfying $T(r, \alpha) = o(1)T(r, g)$ as $r \to \infty$, the following inequality holds:

$$N\left(r, \frac{1}{R(f(z)) - \alpha(z)}\right) \ge (k - 1 + o(1))T(r, f) , \quad r \to \infty .$$

Now we proceed to prove Theorem 4.3.

We will prove the case where $\phi_1(z)$ is transcendental first. (No proof is needed if both ϕ_1 and ϕ_2 are polynomials!).

Let us consider the equation

$$g(U) - z = 0 . \tag{4.17}$$

The equation defines some k-valued algebraic function. Furthermore it is easy to see that the k branches $U_1(z), U_2(z), \ldots, U_k(z)$ are holomorphic in D_U. Letting

$$\phi(z) = f(g(z)) - \phi_1(z) ,$$

and substituting z by $U(z)$ in the above equation, we have

$$\phi(U(z)) = f(g(U(z))) - \phi_1(U(z)) = F(z) - \phi_1(U(z)) , \tag{4.18}$$

where $\phi_1(U(z))$ is an exactly s-valued analytic function $(s \le k)$. This means that the function $\phi_1(U(z))$ can be decomposed in the domain D_U into s distinct single-valued branches in D_U. For an arbitrary complex value a, which is not a Picard exceptional value of $\phi_1(z)$, let $\{\varsigma_j\}$ denote the set of all the roots of the equation $\phi_1(U) - a = 0$ and z_j be the root of $U(z) = \varsigma_j$. It is evident, by Eq. (4.17), that this equation has only one root. If we set $|z_j| = r_j$ and $g(z) = c_o z^k + c_1 z^{k-1} + c_2 z^{k-2} + \ldots + c_k \ (c_0 \ne 0)$. Then

$$r_j = |z_j) = |g(U(z_j)| = |g(\varsigma_j)|$$
$$= (|c_o| + o(1))|\varsigma_j|^k \quad \text{as} \quad j \to \infty \; .$$

This implies that we have the following approximation:

$$|\varsigma_j| = |U(z_j)| = (|c_0| + o(1))^{-\frac{1}{k}} r_j^{\frac{1}{k}} \; , \quad j \to \infty \; . \tag{4.19}$$

We now examine the approximate values of the roots z_j of the equation $\phi_1(U(z)) - a = 0$ in the circle $\{|z| \le r\}$. It is easily verified that for $r \ge r_0$ (where r_0 is some sufficiently large number) any of the root z_j will satisfy the following estimation:

$$|z_j| \le r, \quad |U(z_j)| \le (|c_0| + o(1))^{-\frac{1}{k}} r^{\frac{1}{k}} = A r^{\frac{1}{k}} \quad (j \to \infty) \; , \tag{4.20}$$

where A is a suitable constant.

Due to the one to one correspondence between the points ς_j in $\{|\varsigma| \le A r^{\frac{1}{k}}\}$ and the points z_j in $\{|z| \le r\}$, we have for $r \le r_0$

$$n\left(r, \frac{1}{\phi_1(U(z)) - a}\right) \le n\left(A r^{\frac{1}{k}}, \frac{1}{\phi_1(z) - a}\right) \; .$$

Hence, by virtue of (4.16)

$$N\left(r, \frac{1}{\phi_1(U(z)) - a}\right) \le \frac{k}{s}(1 + o(1)) N\left(A r^{\frac{1}{k}}, \frac{1}{\phi_1(z) - a}\right), r \to \infty \; .$$

Combining Lemma 4.3 and Eq. (4.14), we obtain

$$T(r, \phi_1 \circ U) \le \sum_{\nu=1}^{2s+1} N\left(r, \frac{1}{\phi_1 \circ U - a_\nu}\right) + O(\log r)$$
$$\le \left(\frac{k}{s} + o(1)\right) \sum_{\nu=1}^{2s+1} N\left(A r^{\frac{1}{k}}, \frac{1}{\phi_1(z) - a_\nu}\right) + O(\log r)$$
$$= o(T(A r^{\frac{1}{k}}, f \circ g)), \quad r \to \infty \; . \tag{4.21}$$

Next we prove

$$T(Ar^{\frac{1}{k}}, f \circ g) = O(T(r,f)), \quad r \to \infty .$$

Letting

$$c_o w^k = c_o z^k + c_1 z^{k-1} + \ldots + c_k ,$$

we have

$$w = z \left(1 + \frac{c_1}{c_o z} + \ldots + \frac{c_k}{c_o z^k} \right)^{\frac{1}{k}} .$$

Since the radical tends to 1 as $z \to \infty$, it follows that in the region around ∞, a single-valued branch of the radical function can be selected. Therefore $w(z)$ becomes an analytic function in the domain $\{|z| \geq r_0\}$. Let $Ar^{\frac{1}{k}} > 2r$. The image of the circle $\{|z| = Ar^{\frac{1}{k}}\}$, under the mapping w (that is $1-1$ now), will be some curve γ_r lying in the ring:

$$Ar^{\frac{1}{k}} - d \leq |w| \leq Ar^{\frac{1}{k}} + d ,$$

for some positive constant d. Set

$$f_1(z) = f(c_o z) ,$$

and

$$f_2(z) = f(g(z)) = f(c_o z^k + c_1 z^{k-1} + \ldots + c_k) .$$

Since a point $w_0 \in \gamma_r$ corresponds to each point $z_0 \in \{|z| = Ar^{\frac{1}{k}}\}$ so that $f_2(z_0) = f_1(w_0)$, we deduce

$$T(Ar^{\frac{1}{k}}, f_2) \leq \log M \left(Ar^{\frac{1}{k}}, f_2 \right)$$

$$= \log M(\gamma_r, f_1) \leq \log M(Ar^{\frac{1}{k}} + d, f_1) ,$$

where $M(\gamma_r, f_1) = \max\limits_{z \in \gamma_r} |f_1(z)|$. Using the well-known inequality between $T(r,f)$ and $\log M(r,f)$, we have

$$T(Ar^{\frac{1}{k}}, f(g)) \leq 3T(2(Ar^{\frac{1}{k}} + d), f_1) .$$

From $T(r, f(cz^k)) = T(|c|r^k, f)$ and the fact that the Nevanlinna characteristic function T is an increasing function of r, it follows that

$$T(Ar^{\frac{1}{k}}, f(g)) \leq 3T(|c_o|r^k(Ar^{\frac{1}{k}} + d)^k, f)$$

$$\leq 3T(B_1 r, f), \quad r \to \infty , \qquad (4.22)$$

where B_1 is a suitable positive constant.

On the other hand, we have

$$
\begin{aligned}
T(2Ar^{\frac{1}{k}}, f(g)) &= T(2Ar^{\frac{1}{k}}, f_2) \\
&\geq \frac{1}{3}\log M(Ar^{\frac{1}{k}}, f_2) = \frac{1}{3}\log M(\gamma_r, f_1) \\
&\geq \frac{1}{3}\log M(Ar^{\frac{1}{k}} - d, f_1) \geq \frac{1}{3}T(Ar^{\frac{1}{k}} - d, f_1) \\
&= \frac{1}{3}T(|c_0|(Ar^{\frac{1}{k}} - d)^k, f) \geq \frac{1}{3}T(B_2 r, f) ,
\end{aligned}
\tag{4.23}
$$

where B_2 is a positive constant.

By Eq. (4.14) and letting deg $P = t$, we have

$$
T(r, f(g)) = T(r, F) = (1 + o(1))B_3 r^t, r \to \infty ,
$$

where B_3 is a positive constant. Therefore,

$$
T(Ar^{\frac{1}{k}}, f(g)) = (1 + o(1))B_3 A^t r^{\frac{t}{k}}, \ r \to \infty .
\tag{4.24}
$$

It follows from Eqs. (4.22) and (4.24) that

$$
T(r, f) = O(r^{\frac{t}{k}}) \text{ as } r \to \infty .
$$

It follows from Eqs. (4.22) and (4.23) that

$$
T(Ar^{\frac{1}{k}}, f(g)) = O(r^{\frac{t}{k}}) = O(T(r, f)), \ r \to \infty .
$$

Combining this result and Eq. (4.21) yields

$$
T(r, \phi_1(U)) = o(1)T(r, f) , \quad r \to \infty .
\tag{4.25}
$$

Let α_j be a zero of the function $f(g(U(z))) - \phi_1(z)$, and let $|\alpha_j| = r_j$. Then $t_j = U(\alpha_j)$ will be a zero of $f(g(z)) - \phi_1(z)$. It can easily be verified, from Eqs. (4.19) and (4.20), that $|t_j| \leq Ar^{\frac{1}{k}}$. Therefore

$$
\begin{aligned}
n\left(Ar^{\frac{1}{k}}, \frac{1}{f(g) - \phi_1}\right) &\geq n\left(r, \frac{1}{f(g(U)) - \phi_1(U)}\right) \\
&= n\left(r, \frac{1}{f(z) - \phi_1(U)}\right)
\end{aligned}
$$

and hence

$$N\left(Ar^{\frac{1}{k}}, \frac{1}{f(g(z)) - \phi_1(z)}\right) \geq (1 + o(1))\frac{s}{k}N\left(r, \frac{1}{f - \phi_1(U)}\right) .$$

Thus, by an application of Lemma 4.5, we have

$$N\left(Ar^{\frac{1}{k}}, \frac{1}{f(g) - \phi_1}\right) \geq \left(1 - \frac{1}{s} + o(1)\right)\frac{s}{k}T(r, f), r \to \infty . \qquad (4.26)$$

On the other hand, combining Eqs. (4.13), (4.14), and (4.24) yields

$$N\left(Ar^{\frac{1}{k}}, \frac{1}{f(g) - \phi_1}\right) - N\left(Ar^{\frac{1}{k}}, \frac{1}{\phi_2}\right)$$
$$\leq T(Ar^{\frac{1}{k}}, \phi_2) + O(1) = O(1)T(Ar^{\frac{1}{k}}, f(g))$$
$$= o(1)T(r, f) .$$

Comparing this result with Eq. (4.26), we conclude that $s = 1$. Hence $\phi_1(U(z))$ is, in fact, an (single-valued) entire function. Set

$$\phi_1(U(z)) = \varphi_1(z) . \qquad (4.27)$$

Then

$$\phi_1(U(z)) = \varphi(g(z)) .$$

Choosing a point z_0 such that $g'(z_0) \neq 0$ and $g(z_0) \notin \Gamma_U$, we see that in a neighborhood of the point $w_o = g(z_0)$, there exists a single-valued inverse function $g^{-1}(w)$. Clearly, the function $g^{-1}(w)$ and one branch of the function $U(w)$ are identical in the region of w_0. Let this particular branch be denoted as $u_{j_0}(w)$. Then we have

$$\phi_1(U_{j_0}(g(z))) = \phi_1(z) .$$

Since $\phi_1(U(z))$ is entire, we have, for any $i, j; 1 \leq i, j \leq k$,

$$\phi_1(U_j(g(z))) = \phi_1(U_i(g(z))) = \phi_1(z) .$$

This means that not just in a region around z_0, but the whole z-plane

$$\phi_1(U(g(z))) = \phi_1(z) .$$

From this result and Eq. (4.27), we obtain

$$\phi_1(z) = \varphi_1(g(z)) . \tag{4.28}$$

Let

$$h(\varsigma) = f(\varsigma) - \varphi_1(\varsigma) .$$

Then

$$h(g(z)) = f(g(z)) - \varphi_1(g(z)) = \phi_2(z)e^{P(z)} .$$

By Lemma 4.2, there exists polynomials P_1, P_2 (deg $P_2 <$ deg$P = t$) such that

$$P_1(g(z)) + P_2(z) = P(z) .$$

Consequently

$$h(g(z))e^{-P_1(g(z))} = \phi_2(z)e^{P_2(z)} .$$

Note that the left-hand side of the above equation is an entire function of g, that is,

$$\phi_2(z)e^{P_2(z)} = \varphi_2(g(z)) ; \quad \varphi_2(\varsigma) = h(\varsigma)e^{-P_1(\varsigma)} . \tag{4.29}$$

We have therefore proved that

$$f(g(z)) = \varphi_1(g(z)) + \varphi_2(g(z))e^{P_1(g(z))} .$$

In view of this result and Eqs. (4.28) and (4.29), we see that the theorem is proven for the case where $\phi_1(z)$ is a transcendental entire function. As a result of Theorem 4.1 we now only need to deal with the case where $\phi_1(z)$ is a polynomial and $\phi_2(z)$ is a transcendental entire function. In this situation, we express $P(z)$ as $P(z) = P_1(g(z)) + P_2(z)$, deg $P_2 <$ deg P and consider

$$h(g(z)) = f(g(z))e^{-P_1(g(z))} = \phi_2(z)e^{P_2(z)} + \phi_1(z)e^{-P_1(g(z))} . \tag{4.30}$$

Only two case are possible:

Case (i). $\phi_2(z)e^{P_2(z)}$ is a transcendental entire function. In this case note that $T(r, \phi_2 e^{P_2}) = \circ T(r, h(g))$ and we arrive at the case that has been discussed above.

Case (ii). $\phi_2(z)e^{P_2(z)}$ is a polynomial. Theorem 4.1 is now applicable. Theorem 4.3 is thereby proven.

Recall that Goldstein proved (in Theorem 4.2) that if Eq. (4.14) in Theorem 4.3 is strengthened by requiring that the orders of ϕ_1 and ϕ_2 be less than the degree of P, then the factorization $F(z) = f(g(z))$ for any two nonlinear functions f and g will lead to the conclusion that g must be a polynomial.

Incidentally, while Goldstein and Prokopovich obtained their results, Gross and Yang derived a result similar to that of Goldstein and Prokopovich's. Their method was entirely different in that it utilizes Theorem 4.2, and some elementary properties of an algebraic function and its inverse function.

Theorem 4.4. (Gross and Yang) Let $P(z)$ be polynomial of degree $m(\geq 1)$, and let $h(z)(\not\equiv 0)$ and $k(z)(\not\equiv \text{constant})$ be two entire functions of order less than m. Then $h(z)e^{P(z)} + k(z)$ is either prime or it can only be factorized as

$$h(z)e^{P(z)} + k(z) = f(L(z)) , \qquad (4.31)$$

where $L(z)$ is a nonlinear polynomial of degree n; $f(z) = \mu(z)\exp[cz^d] + \beta(z)$ is an entire function; α and β are entire functions of order less than m; and c is a constant $\neq 0$. Furthermore the following three relationships are satisfied:

(i) $n|m$ (namely $\frac{m}{n}$ is an integer),
(ii) $h(z)e^{P(z)} \equiv \alpha(L(z))\exp[cL(z)^d], d = \frac{m}{n}$,
(iii) $k(z) = \beta(L(z))$.

Before going into the proof of the theorem, we need the following two lemmas.

Lemma 4.6. Let $P(z,w) = z^n + a_1 z^{n-1} + a_2 z^{n-2} + \ldots + a_n - w$, where a_i $(i = 1, 2, \ldots, n)$ are constants, and write $P(z,w) = \prod_{i\equiv 1}^{n}(z - z_i(w))$. Then every elementary symmetric function in $z_i(w)$ is a polynomial in w.

Proof. It is obvious.

Lemma 4.7. Let $w, z_i(w)(i = 1, 2, \ldots, n)$ be as in Lemma 4.6 and $g(z)$ be an entire function. Then $\sum_{i=1}^{n} g(z_i(w))$ is an entire function in w.

Proof. Let us express g in its Taylor series:

$$g(z) = a_0 + a_1 z + a_2 z^2 + \ldots = \sum_{j=0}^{\infty} a_j z^j .$$

Then in any bounded domain D we have

$$\sum_{i=1}^{n} g(z_i(w)) = \sum_{i=1}^{n} \sum_{j=0}^{\infty} a_j [z_i(w)]^j \ .$$

Since the infinite series is convergent absolutely in every bounded domain, we can rearrange the double summation and obtain

$$\sum_{i=1}^{n} g(z_i(w)) = \sum_{j=0}^{\infty} \left(\sum_{i=1}^{n} a_j [z_i(w)]^j \right) = \sum_{j=0}^{\infty} P_j(w) \ ,$$

where the P_j $(j = 1, 2, \ldots)$ are polynomials. This proves the lemma.

Proof of Theorem 4.4. Suppose that $F = he^P + k$ is not prime. By Theorem 4.2 the only possible factorization for F is of the form:

$$he^P + k = f(L) \ ,$$

where f is entire and

$$L(z) \equiv a_0 + a_1 z + \ldots + a_n z^n, \quad a_i \text{ constants}, \quad (a_n \neq 0) \ .$$

Let

$$P(z) = c_0 + c_1 z + \ldots + c_m z^m \quad (c_m \neq 0) \tag{4.32}$$

and

$$w = L(z) \ .$$

We have, for sufficiently large $|w|$ (it will be assumed without loss of generality that all w's below are sufficiently large):

$$L(z) - w \equiv a_n(z - z_1(w))(z - z_2(w)) \ldots (z - z_n(w)) \ , \tag{4.33}$$

where $z_i(w) (i = 1, 2, \ldots, n)$ are n-distinct branches (since, clearly, $P(z, w) \equiv L(z) - w$ is irreducible and of degree n in z). For $i = 1, 2, \ldots, n$ we have an expansion of the form

$$z_i(w) = w_i^{\frac{1}{n}} (b_0 + b_{-1} w^{-\frac{1}{n}} + b_{-2} w^{-\frac{2}{n}} + \ldots) \tag{4.34}$$

that is valid in ∞, where $w_i^{\frac{1}{n}} = \rho_i w_1^{\frac{1}{n}}, \rho_i$ $(i = 1, 2, \ldots, n)$ are n-distinct roots of unity, $\rho_1 = 1$, and $w_1^{\frac{1}{n}}$ is a fixed branch of $w^{\frac{1}{n}}$. For $z_i(w)$ as above, we have

$$h(z_i(w)) \exp(P(z_i(w))) + k(z_i(w)) = f(w) \ , \quad i = 1, 2, \ldots, n \ . \tag{4.35}$$

Thus, for $i \neq 1$, we obtain the following result

$$\frac{h(z_i(w))\exp(P(z_i(w))) + k(z_i(w))}{h(z_1(w))\exp(P(z_1(w))) + k(z_1(w))} = 1 . \tag{4.36}$$

After dividing through the denominator and the numerator by $h(z_1(w)) \times \exp(p(z_i(w)))$, the above quotient becomes

$$\frac{h(z_i(w))/h(z_1(w))\exp(P(z_i(w))}{1 + k(z_1(w))/h(z_1(w))\exp(-P(z_1(w)))}$$

$$-\frac{P(z_1(w)) - k(z_i(w)/h(z_1(w)))\exp(-P(z(w)))}{1 + k(z_1(w))/h(z_1(w))\exp(-P(z_1(w)))} . \tag{4.37}$$

If we substitute Eq. (4.34) into Eq. (4.32), then

$$\begin{aligned}
P(z_i(w)) &= c_m[w_i^{\frac{1}{n}}(b_0 + b_{-1}w_i^{-\frac{1}{n}} + b_{-2}w_i^{-\frac{2}{n}} + \ldots)]^m \\
&\quad + c_{m-1}[w_i^{\frac{1}{n}}(b_0 + b_{-1}w_i^{-\frac{1}{n}} + b_{-2}w_i^{-\frac{2}{n}} + \ldots)]^{m-1} \\
&\quad + \ldots + c_0 \\
&= d_m w_i^{m/n} + d_{m-1}w_i^{(m-1)/n} + \ldots \\
&\quad + d_0 + d_{-1}w_i^{-\frac{1}{n}} + d_{-2}w_i^{-\frac{1}{2n}} + \ldots .
\end{aligned} \tag{4.38}$$

We are going to verify that $n|m$ by assuming the contrary $n \nmid m$. Next, we choose a path of w, l, a straight line in the w-plane tending to infinity such that along l, $d_m w_1^{m/n} = |d_m w_1^{m/n}|$.

Now from Eq. (4.38) we have

$$P(z_i(w)) - P(z_1(w)) = d_m(\rho_i^m - 1)w_1^{m/n} + d_{m-1}(\rho_i^{m-1} - 1)w_1^{(m-1)/n} + \ldots . \tag{4.39}$$

If $n|m$, then

$$\mathrm{Re}(\rho_1^m - 1) < 0 \quad (i \neq 1) . \tag{4.40}$$

We now examine the behavior of the functions $k(z_1(w))/h(z_1(w))$, and $k(z_i(w))/h(z_1(w))$ along l. By assuming that h is an entire function of order less than m and by the property of the minimum modulus, we have for any given $\varepsilon > 0$ and sufficiently large r, $m_h(r) \neq O(1)\exp(-r^{m-\varepsilon})$, where $m_h(r) = \min_{|z|=r}|h(re^{i\theta})|$. From this result, noting Eqs. (4.39), (4.40), and that $k(z)$ is an entire function of order less than m, we see that

$|\exp(P(z_i(w))) - \exp(P(z_1(w)))|$ grows much slower than $\exp(-\alpha|w|^{m/n})$ as $w \to \infty$ for some constant $\alpha > 0$. We conclude, after a simple verification, that the following three quantities:

$$[h(z_i(w))/h(z_1(w))]\exp(-P(z,(w)))\,,$$

$$[k(z,(w))/h(z_1(w))]\exp(-P(z_1(w)))\,,$$

and

$$[k(z_i(w))/h(z_1(w))]\exp(-P(z_1(w)))$$

all tend to 0 as $w \to \infty$ through a suitable sequence $\{w_n\}$ on l. Therefore the left hand of Eq. (4.36) tends to 0 as $w \to \infty$ which is a contradiction. Hence, we must have $n|m$. Therefore, from Eq. (4.37), we have

$$P(z_j(w)) = d_m w^d + d_{m-1} w_i^{(m-1)/d} + \dots\,,$$

where $d = m/n$ is an integer. Substituting this into Eq. (4.19) for $i = 1, 2, \dots, n$ and adding, we obtain

$$\exp(d_m w^d)\left[\sum_{i=1}^{n} h(z_i(w))\exp(d_{m-1} w_1^{(m-1)/n} + \dots)\right]$$
$$+ \sum_{i=1}^{n} k(z_i(w)) = nf(w)\,. \tag{4.41}$$

Now, according to Lemma 4.7, $\sum_{i=1}^{n} k(z_i(w)) \equiv T(w)$ is an entire function of order less than d (since it is easy to verify that each function $k(z_i(w))$ grows no faster than $e^{r^{d-\epsilon_i}}$; where ϵ is small positive constant). It is also easy to verify that the growth of $h(z_i(w))$ (for $i = 1, 2, \dots, n$) is no faster than that of $e^{r^{d-\delta_i}}$; where δ_i is a small positive constant. We also note that the function $\exp[d_{m-1} w^{(m-1)/n} + \dots]$ grows no faster than $e^{r^{-\eta_i}}$; where η_i is a positive constant (for $i = 1, 2, \dots, n$), when $|w|$ is sufficiently large.

Equation (4.41) can be rewritten as:

$$\sum_{i=1}^{n} h(z_i(w))\exp(d_{m-1} w_i^{(m-1)/n} + \dots) = [nf(w) - T(w)]\exp(-d_m w^d)\,.$$
$$\tag{4.42}$$

Clearly, the right-hand side of Eq. (4.42) is an entire function. Furthermore, the left-hand side by virtue of the above estimates grows no faster than

$e^{r^{d-\eta}}$ for sufficiently large r; where η is a positive constant $< d$. Thus, we conclude that $S(z) \equiv [nf(w) - T(w)] \exp(-d_m w^d)$ is an entire function of order less than d. Consequently,

$$f(w) \equiv (S(w)/n) \exp(d_m w^d) + T(w)/n .$$

From this result and Eq. (4.31), we have

$$h(z) \exp(P(z)) + k(z) \equiv (S(L(z))/n) \exp(d_m (L(z))^d) + T(L(z))/n .$$

Hence

$$h(z) \exp(P(z)) - \frac{S(L(z)) \exp(d_m (L(z))^d)}{n} \equiv -k(z) + \frac{T(L(z))}{n} . \qquad (4.43)$$

Since $T(w)$ and $S(w)$ are entire functions in w of order less than d $(= m/n)$ and $L(z)$ is a polynomial of degree n, it follows that both $T(L(z))$ and $S(L(z))$ are entire functions in z of oder less than m. Thus the right-hand side of Eq. (4.43) is an entire function of order less than m. Now the left-hand side can be expressed as

$$\exp(P(z))[(h(z) - S(L(z))/n) \exp(d_m (L(z))^d - P(z))] .$$

To show that $(h(z) - S(L(z))/n) \exp(d_m (L(z))^d - P(z)) \equiv U(z) \equiv 0$ we will suppose the contrary $(U(z) \not\equiv 0)$ to be true. This means that $k(z) - T(L(z))/n \not\equiv 0$. If $d_m L(z)^d - P(z) \equiv q(z)$ has a degree $= m$, then the function $(S(L(z))/n) \exp(d_m (L(z)) - P(z))$ will have three functions: $\infty, 0$, and $h(z)$ as its deficient functions. This is impossible. Alternatively, $q(z)$ can only be a polynomial of degree less than m. This implies that $U(z)$ is of an order less than m. Therefore, the order of $e^{P(z)} U(z)$ is m (since $k(z) - T(L(z))/n$ has an order less than m) which results in a contradiction. We must conclude

$$U(z) \equiv 0 ,$$

and hence, from Eq. (4.43),

$$k(z) - T(L(z))/n \equiv 0$$

or

$$k(z) \equiv T(L(z))/n .$$

Consequently we have

$$h(z)\exp(P(z)) \equiv (S(L(z))/n)\exp(d_m L(z)^d) \, .$$

Theorem 4.4 is proven.

Question. Why is the assumption $k(z) \not\equiv$ constant is necessary in the theorem? (Exhibit a counter example!)

In the proof of Theorem 4.4 we used Theorem 4.2 to show that $F(z) = he^P + k$ is pseudo-prime and that a factorization of the form $F = q(f)$ is impossible if q is a nonlinear polynomial and f is entire.

Theorem 4.6 in the next section will also yield the result that $he^P + k$ is pseudo-prime. Application of the Tumura-Clunie theorem, allows for exclusion of the factorization of the form $F = q(f)$. Rewriting $F = he^P + k = q(f)$ as $he^P = q(f) - k$, leads to the impossible conclusion $c_o(f - c_1)^n \equiv he^P$; where c_o is the leading coefficient of $q(z)$, $n = \deg q$, and c is a constant.

In view of the above results, we draw the following conclusion.

Corollary 4.2. Let $P_i(z)(i = 1, 2, \ldots, m)$ denote a polynomial of degree t_i and let $h_i(z)(i = 1, 2, \ldots, m)$ denote a non-constant entire function of order less than t_i. Assume that $0 \leq t_1 < t_2 < \ldots < \ldots < t_m; m \geq 2$. Then $F(z) \equiv \sum^m h_i(z)e^{P_i(z)}$ is pseudo-prime. Moreover, if F is not prime, then the only possible form of the factorization of F is $F = f(g)$, where f is entire and q is a nonlinear polynomial and a common right factor for all of the terms $h_i e^{P_i}, i = 1, 2, \ldots, m$.

4.4. THE CRITERIA OF PSEUDO-PRIMENESS FOR ENTIRE FUNCTIONS

To simplify the proof, we shall mainly deal with entire functions. In general, the steps to prove that a given transcendental entire function F is prime are (i) first prove that F is pseudo-prime (ii) prove that F cannot be expressed as $F(z) = P(g(z))$, where F is entire and P is a polynomial with deg $P \geq 2$, and (iii) prove that F cannot be expressed as $F(z) = h(g(z))$; where $q(z)$ is a polynomial of degree ≥ 2 and h an entire function.

Now we introduce some sufficient conditions for determining the pseudo-primeness of a given entire function.

Theorem 4.5. (Goldstein) Let $F(z)$ be a finite-ordered entire function with $\delta(a, F) = 1$ for some complex number a. Then F is pseudo-prime.

(The function $F(z) = e^{e^z}$ shows that the restriction on the order of F is a necessary one).

Proof. We may assume, without loss of generality, that $a = 0$. By the assumption we have

$$K(F) = \overline{\lim_{r \to \infty}} \{N(r, 0, F) + N(r, \infty, F)\} / T(r, F)$$
$$\leq \overline{\lim_{r \to \infty}} \frac{N(r, 0, F)}{T(r, F)} + \overline{\lim_{r \to \infty}} \frac{N(r, \infty, F)}{T(r, F)}$$
$$= 1 - \delta(0, F) + 1 - \delta(\infty, F) = 0 . \tag{4.44}$$

Using a result of Edrei-Fuchs' (*Comm. Math. Helv.* **33** (1959), f.4, 258-295), it follows that the lower order of F satisfies $p - \frac{1}{2} \leq \mu < p + \frac{1}{2}$ for some positive integer p. In this case the following result holds: To any finite ordered entire function F satisfying conditions of Eq. (4.44) there exists a sequence of circular arcs $\{\gamma_j\}_{j=1}^{\infty}$, i.e., γ_j lying on the circle $\{|z| = r_j\}$ with $r_j \uparrow \infty$. Moreover, the angular measurement of γ_j is $\geq 2\pi/3p$ for $j = 1, 2, \ldots$ so that when $z \in A(= \cup \gamma_j)$ and $|z| \geq r_0$ (r_0 is sufficiently large) the following estimation holds:

$$\log |F(z)| \leq -\frac{\pi}{16} T(r, F) . \tag{4.45}$$

In the same paper, Edrei-Fuchs also proved that these exists a sequence of segments $\{l_j\}_{j=1}^{\infty}$ on which F also satisfies the inequality in Eq. (4.45), where l_j defines a segment with one end point as $r_j e^{i\theta_j} \in \gamma_j$ and the other end point as $r_{j+1} e^{i\theta_{j+1}} \in \gamma_{j+1}, j = 1, 2, \ldots$. Therefore $\{\gamma_j\}_{j=1}^{\infty}$ and $\{l_j\}_{j=1}^{\infty}$ form a continuous curve L which Eq. (4.45) holds when $|z|$ is sufficiently large.

If F were not pseudo-prime, then we would have $F(z) = f(g(z))$ where both f and g are trancendental entire functions. According to a theorem of Polya's (noting that F has finite order), f is of zero order. Hence, there exists a real sequence $\{R_n\}$ with $R_n \uparrow \infty$ such that

$$\min_{|z|=R_n} |f(z)| \to \infty \quad \text{as } n \to \infty . \tag{4.46}$$

But by Eq. (4.45) $F(z) \to 0$ as z tends to ∞ along L. Therefore we must conclude that $\Gamma = g(L)$, that is, the image of L under g is a bounded curve, otherwise, Eqs. (4.45) and (4.46) will yield a contradiction. Next we show that $g(z) \to \alpha$ when z tends to ∞ along L, where α is a zero of $f(z)$. Let $z = h(t)(0 \le t < \infty)$ be the equation of the curve L, $h(t) \to \infty$ as $t \to \infty$. Then Γ can be represented as

$$U = g(z) = g(h(t)) , \quad 0 \le t < \infty .$$

Since Γ is bounded, there exists some closed disc $K = \{z||z| \le R\}$ containing Γ. Without loss of generality, we may assume that f does not vanish on $|z| = R$. Let $\alpha_j, j = 1, 2, \ldots, k$, be all the distinct zeros of f in $|z| < R$ and $\delta = \min_{i,j=1,2,\ldots,k \atop i \ne k}\{d_j, |\alpha_i - \alpha_j|\}$ where d_i defines the distance from α_i to the complement set of K. Let m be any positive number and C_i denote the circle $|z - \alpha_i| = \delta/m$. Letting $\min_{z \in C_i} |f(z)| = m_i (> 0)$ and $m_0 = \min\{m_1, m_2, \ldots, m_k, \min_{z \in R} |f(z)|\}$, we have $m_0 > 0$. Therefore, $|f(z)| < m_0$ for some $z \in K$, implies that the point z must lie outside the circles $C_i, i = 1, 2, \ldots, k$. When $t \ge t_0$ then $g(h(t)) \in K$ and $|f(g(h(\varepsilon)))| < m$. This means that when $t \ge t_0, g(h(t))$ lies inside some circle C_i. The set $\{g(h(t))|t \ge t_o\}$ is connected. It follows from this and the fact that the sets $C_i(i = 1, 2, \ldots, k)$ are mutually separated that there exists some positive integer, $j(1 \le j \le k)$ such that the following inequality holds:

$$|g(h(t)) - \alpha_j| < \frac{\delta}{m}, \quad (t \ge t_0) .$$

Since m can be arbitrarily large, this implies that $g(z) \to \alpha$ as $z \to \infty$ along L, where α is a zero of f. Therefore we have, given any $\varepsilon > 0$ and $z \in L$ with $|z| \ge r_0$,

$$|g(z) - \alpha| \le \varepsilon, \quad \forall z \in \gamma_i, \quad j \ge j_0 . \tag{4.47}$$

Assume that the multiplicity of the zero point α is $s(\ge 1)$. It follows that there exists a positive constant A (> 0) such that whenever $|z - \alpha| \le \varepsilon$,

$$|f(z)| \ge A|z - \alpha|^s , \tag{4.48}$$

that is, whenever $|g(z) - \alpha| \le \varepsilon$,

$$|f(g(z))| \ge A|g(z) - \alpha|^s . \tag{4.49}$$

Thus, for $z \in \gamma_i, j \ge j_0$ (or $z \in L, |z| \ge r_0$)

$$|F(z)| \ge A|g(z) - \alpha|^s . \tag{4.50}$$

Since the inequality in Eq. (4.45) holds for $z \in \gamma_j$, we obtain

$$s \log |g(z) - \alpha| + \log A \leq -\frac{\pi}{16} T(|z|, F) .$$

Consequently, for $z \in \gamma_j (j \geq j_0)$, we have

$$\log^+ \left| \frac{1}{g(z) - \alpha} \right| \geq \log \left| \frac{1}{g(z) - \alpha} \right| \geq \frac{\pi}{16s} T(|z|, F) + \frac{\log A}{s} .$$

When $j \geq j_0$ with $z = re^{i\theta}$ from the above results and by applying Nevanlinna first fundamental theorem, we obtain, for an integer $p > 0$

$$T(r_j, g) + O(1) \geq m(r_i, \alpha, g) = \frac{1}{2\pi} \int_0^{2\pi} \log^+ \left| \frac{1}{g(re^{i\theta}) - \alpha} \right| d\theta$$

$$\geq \frac{1}{2\pi} \int_{\gamma_j} \log^+ \left| \frac{1}{g(re^{i\theta}) - \alpha} \right| d\theta$$

$$\geq \frac{1}{3p} \left(\frac{\pi}{16s} T(r_j, F) + \frac{\log A}{s} \right) . \tag{4.51}$$

Since $T(r, F) \to \infty$ as $r \to \infty$, the above equation yields

$$T(r_j, g) \geq B T(r_j, F), \quad j \geq j_0 , \tag{4.52}$$

where B is a suitable positive constant. However, according to a theorem of Clunie's, for any two transcendental entire functions g and f,

$$\lim_{r \to \infty} \frac{T(r, f(g))}{T(r, g)} = \infty . \tag{4.53}$$

This contradicts with Eq. (4.52). We must conclude that it is impossible for both f and g to be transcendental entire in the factorization $F = f(g)$. This completes the proof.

Remark. Goldstein remarked that Theorem 4.5 remains valid under either of the following two conditions:

(i) $\delta(0, F') = 1$

(ii) $\sum_{a \neq \infty} \delta(a, F) = 1$.

It was also remarked that the Edrei-Fuchs' result applies not only for $\delta(a, F) = 1$ but also for $\delta(a, F) > 1 - \varepsilon(\rho)$, where $\varepsilon(\rho)$ is a positive constant

$(0 < \varepsilon(\rho) < 1)$ depending on the order of F. The above remarks also lead to an interesting conjecture as follows.

Conjecture 2. (Fuchs) Let F be an entire function of finite order. If $\delta(a, F) > 0$ for some complex number, then F is pseudo-prime.

Using an argument similar to that used for the preceding theorem the following result can be obtained.

Theorem 4.6. (Gross and Yang) Let $P(z)$ be a polynomial of degree t (≥ 1) and $h_1(z)$ and $h_2(z)(\not\equiv 0)$ be two entire functions of order less than t. Then $F(z) \equiv h_1(z)e^{P(z)} + h_2(z)$ is pseudo-prime.
Hint: Write $F(z)$ as $h_2(z)\{\frac{h_1(z)}{h_2(z)}e^{P(z)} + 1\}$.

Question. Does the theorem remain valid if only $T(r, h_i) = o(1)T(r, e^P)$ $r \to \infty, i = 1, 2$ is assumed?

Recall that a transcendental entire function F is called left-prime or E-left-prime if $F = f(g)$ with f and g being entire implies that f must be linear whenever g is transcendental. F is called right-prime or E-right-prime if $F = f(g)$ with f and g being entire implies g must be linear whenever f is transcendental. Clearly we have

(i) If E is both right and left-prime, then F is E-prime.
(ii) A left or right-prime transcendental entire function must be a pseudo-prime.

We now provide some criteria for left-primeness.

Theorem 4.7. (Ozawa) Let $F(z)$ be an entire function of finite order whose derivative $F'(z)$ has an infinite number of zero. Suppose for any complex number c, the following simultaneous equations:

$$\begin{cases} F(z) = c \\ F'(z) = 0 \end{cases} \tag{4.54}$$

have only a finite number of solutions. Then $F(z)$ is left-prime.

Proof. Suppose that F has the factorization $F = f(g)$; with f and g being transcendental entire functions. From Polya's theorem we must have $\rho(F) = \rho(F') = 0$. Hence $f'(\varsigma)$ has an infinite number of zero, that can be summarized as $\{\varsigma_j\}_{j=1}^{\infty}$. There must be some fixed ς_j such that the solution to the equation $g(z) = \varsigma_j$ are an infinite set. Let $\{z_n\}_{n=1}^{\infty}$ be the set. The simultaneous equations

$$\begin{cases} F(z_n) = f(g(z_n)) = f(\varsigma_j) = c \\ F'(z_n) = f'(g(z_n))g'(z_n) = 0, \quad n = 1, 2, \ldots \end{cases}$$

have an infinite number of solutions. This is a contradiction to the hypothesis. We conclude that F must be pseudo-prime.

Assume that $F = P(g)$, where P is a nonlinear polynomial and g is an entire function. $P'(\varsigma)$ has at least one zero, ς. If $g(z) = \alpha$ results in an infinite number of solutions, then using the same argument as above we will get a contradiction. Now assume that $g(z) = \alpha$ only has a finite number of solutions, this results in

$$g(z) = \alpha + Q(z)e^{q(z)} ,$$

where q and Q are polynomials. This gives $g'(z)$ a finite number of zeros. Since the assumption states that $F'(z) = P'(g(z))g'(z)$ has an infinite number of zeros, it follows that there must exist a root of $P'(\varsigma)$, β, not equal to α, such that $g(z) = \beta$ has an infinite number of solutions. This will again lead to a contradiction. The theorem is thus proved.

Exercise. Prove that $F(z) = e^z + P(z)$, where P is a polynomial, is left-prime. Use this result to show that F is E-prime.

Exercise. Illustrate the requirement that $F'(z)$ has an infinite number of zeros is a necessary condition for the validity of Theorem 4.7.

When no restriction is imposed on the order of $F(z)$, the following results.

Theorem 4.8. (Ozawa) Let $F(z)$ be a transcendental entire function with $N\left(r, \frac{1}{F'}\right) \geq kT(r, F)$ for some positive constant k. If for any complex number c, the system of Eqs. (4.54) has only a finite number of solutions, then F is left-prime.

Proof. Suppose that $F = f(g)$, where f and g are both transcendental entire. Finally we assume that $f'(\varsigma) = 0$ has no roots at all. Then

$$N\left(r, \frac{1}{F'}\right) = N\left(r, \frac{1}{g'}\right) \leq T(r, g') + \mathrm{O}(1) \leq (1 + \varepsilon)T(r, g), \text{n.e.,} \quad (4.55)$$

where "n.e." means the inequality holds nearly everywhere for sufficiently large values of r except possibly a set of r values of finite length.

On the other hand, for any positive integer p and some constant A (not a Picard exceptional value of f), we have

$$T(r, F) \geq N\left(r, \frac{1}{F - A}\right) + O(1) \geq \sum_{j=1}^{p} N\left(r, \frac{1}{g - \alpha_j}\right) + O(1),$$

$$\geq (p - 1)T(r, g) + O(\log rT(r, g)) \text{ n.e.}, \tag{4.56}$$

where $\alpha_j \in f^{-1}(A)$.

The combination of Eqs. (4.55) and (4.56) yields a result that will contradict the hypothesis of the theorem: $N\left(r, \frac{1}{F'}\right) \geq kT(r, F)$. If we assume that f has only one zero, ς_0, and $g(z) = \varsigma_0$ has a finite number of roots, it follows that

$$N\left(r, \frac{1}{F'}\right) = N\left(r, \frac{1}{g'}\right) = O(\log r) \leq T(r, g') + O(\log r)$$

$$\leq T(r, g) + O(\log rT(r, g)) \quad \text{n.e.} .$$

This leads to the same contradiction found in the previous case. Alternatively we assume that $f'(\varsigma)$ has only one zero, ς_0, but $g(z) = \varsigma_0$ has an infinite number of roots, $\{z_j\}$. Then the following simultaneous equations

$$\begin{cases} F(z) = f(\varsigma_0) \\ F'(z) = 0 \end{cases}$$

have an infinite number of solutions $\{z_j\}$. This is also a contradiction to the hypothesis.

Now we assume that $f'(\varsigma)$ has at least two distinct zeros. By choosing one of the roots, ς_1 so that $g(z) = \varsigma_1$ has an infinite number of roots, we will arrive at the same contradiction. This also proves that F is a E-pseudo-prime. Finally, we assume that $F = P(g)$, where P is a polynomial and g is a transcendental entire function. If $P'(\varsigma)$ has only one zero and $g(\varsigma) = \varsigma_0$ has a finite number of roots, then

$$g(z) = \varsigma_0 + Q(z)e^{G(z)}, \quad g'(z) = (Q' + G'Q)e^{G(z)}$$

where Q is a polynomial and G is an entire function. Then

$$N\left(r, \frac{1}{g'}\right) \leq N\left(r, \frac{1}{Q' + G'Q}\right) = \circ T(r, g) \quad \text{n.e.} . \tag{4.57}$$

If on the other hand, $t = \deg P$, then there exist some arbitrarily small positive number ε and ε' such that

$$N\left(r, \frac{1}{g'}\right) = N\left(r, \frac{1}{F'}\right) + O(\log r)$$

$$\geq (1+\varepsilon)kT(r, F) \geq k(t-1)(1+\varepsilon')T(r, g) .$$

This will contradict Eq. (4.57) unless $t = 1$, i.e., P is a linear polynomial. The cases, like $P'(\varsigma) = 0$ can be proposed as having a root, ς_0, such that $g(z) = \varsigma_0$ has an infinite number of roots, or $P'(\varsigma)$ can have at least two distinct zeros and one of them, ς_1, can enable $g(z) = \varsigma_1$ to have an infinite number of roots, etc.; can be argued as before and similar contradictions will result. This also completes the proof of the theorem.

Remark. Noda noted that the condition $N\left(r, \frac{1}{F'}\right) > kT(r, F)$ of the theorem can be replace by either (i) requiring $N\left(r, \frac{1}{F'}\right) \geq kT(r, F)$ on a set of r values of infinite measure for some $k > 0$ or (ii)

$$N\left(r, \frac{1}{F'}\right) - \left[N\left(r, \frac{1}{F}\right) - \overline{N}\left(r, \frac{1}{F}\right)\right] \geq kT\left(r, \frac{F'}{F}\right), \quad \text{n.e.} .$$

These two facts are useful in the proof of Theorem 4.9 found in the next section.

4.5. THE DISTRIBUTION OF THE PRIME FUNCTIONS

We would like to know like the distribution of prime number r in the set of integers; the distribution of prime functions in the family of entire functions. In the section we shall resolve two related questions:

(A) (Gross) Given any entire function f, does there exist a polynomial Q such that $f + Q$ is prime?

B) (Gross, Osgood and Yang) Given any entire function f, does there exist an entire function g such that gf (the product) is prime?

Noda provided affirmative answers to the above two questions as follows.

Theorem 4.9. (Noda) Let $f(z)$ be a transcendental entire function. Then the set $\{a|a \in \mathbb{C} \text{ and } f(z) + az \text{ is not prime}\}$ is at most a countable set in the complex plane $\mathbb{C}$.

We shall first prove:

Lemma 4.8. Let $f(z)$ be a transcendental entire function. Then there is a countable set E of complex numbers such that the simultaneous equations

$$\begin{cases} f(z) - az = c \\ f'(z) - a = 0 \end{cases} \tag{4.58}$$

have no more than one common root for any constant $c(\in \mathbb{C})$ provided that a belongs to $\mathbb{C}/E$.

Proof. We write

$$A = \mathbb{C} \backslash \{z \in \mathbb{C} | f''(z) = 0\} \; .$$

Clearly A is an open set. It is easily verified that one can choose an open covering of A, say $\{C_i\}_{i=1}^{\infty}$, such that the following three conditions will be satisfied:

(i) $\bigcup_{i=1}^{\infty} C_i = A$,

(ii) $f'(z)$ is univalent in $C_i (i = 1, 2, \dots)$ and

(iii) $D_i = \{f'(z) | z \in C_i\}$ is a disk $(i = 1, 2, \dots)$.

Set

$$F(z) = f(z) - z f'(z) \; . \tag{4.59}$$

and define functions U_i and $V_i (i = 1, 2, \dots)$ as follows:

$$U_i(w) = (f' | C_i)^{-1}(w) \quad (w \in D_i \; , \quad i = 1, 2, \dots); \tag{4.60}$$

namely $f \circ U_i(w) = w \; \forall w \in D_i$,

$$V_i(w) = F(U_i(w)) \quad (w \in D_i, \quad i = 1, 2, \dots) \; . \tag{4.61}$$

Let

$$I = \{(i, j) | \text{ with } i, j \text{ positive integers such that } D_i \cap D_j \neq \phi \; , \\ \text{and } V_i(w) \neq V_j(w) \; (w \in D_i \cap D_j)\} \tag{4.62}$$

$$S_{sj} = \{w | w \in D_i \cap D_j, \; (i, j) \in I \text{ and } V_i(w) = V_j(w)\} \; ,$$

and

$$E_0 = \left(\bigcup_{i=1}^{\infty} D_i \right) \backslash \left(\{f'(z); f''(z) = 0, z \in \mathbb{C}\} \cup \left\{ \bigcup_{(i,j) \in I} S_{ij} \right\} \right) \; . \tag{4.63}$$

It becomes easy to verify that $E = \mathbb{C} \backslash E_0$ is at most a countable set.

Let $a \in E_0$. By the definition of the set E_0, if there exists a pair of integres $i, j \, (i \neq j)$ such that

$$V_i(a) = V_j(a) , \tag{4.64}$$

then by (4.62) and (4.63)

$$V_i(w) \equiv V_j(w) \quad \forall w \in D_i \cap D_j (\neq \phi) .$$

Thus

$$V_i'(a) = V_j'(a) .$$

From Eqs. (4.59), (4.60), and (4.61), we obtain

$$V_i'(a) = -U_i(a), \quad V_j'(a) = -U_j(a) .$$

This gives us

$$U_i(a) = U_j(a) . \tag{4.65}$$

Again Eqs. (4.59), (4.60), and (4.61) we have

$$V_i(a) = f(U_i(a)) - aU_i(a), \quad V_j(a) = f(U_j(a)) - aU_j(a) .$$

From (4.60) and (4.64) we see that if

$$f(U_i(a)) - aU_i(a) = f(U_j(a)) - aU_j(a) ,$$

then

$$U_i(a) = U_j(a) .$$

On the other hand, by Eq. (4.60) and (4.63), we see that

$$D \equiv \{U_k(a) | a \in D_k; \ k = 1, 2, \ldots\} = \{z_n | f'(z_n) = a, n = 1, 2, \ldots\} .$$

This means that the set coincides with the set of distinct a-points $\{z_n\}$ of $f'(z)$. Therefore, if $z_n \neq z_m$, then $f(z_n) - az_n \neq f(z_m) - az_m$. Thus, for any $a \in E_0$, the simultaneous equations

$$\begin{cases} f(z) - az = c \\ f'(z) - a = 0 \end{cases}$$

have at most one common root for any constant c. This also concludes the proof of Lemma 4.8.

Proof of Theorem 4.9. Let $k \in (0, \frac{1}{2})$. It follows from Lemma 4.8 and the second fundamental theorem that there exists a countable set E_1 of complex numbers such that the conclusion of Lemma 4.8 remains valid with E replaced by E_1 and that for every $a \in \mathbb{C} \backslash E_1$

$$N\left(r, \frac{1}{f' - a}\right) \geq kT(r, f), \qquad (4.66)$$

is valid on a set (depending on a) of 2 values of infinite measure.

We have by Lemma 4.8, that $f(z) - az$ is left-prime provided $a \in \mathbb{C} \backslash E_1$. Next we shall show that $f(z) - az$ is right-prime (in entire sense).

Let $f(z) - az = g(P(z))$, where g is transcendental entire and P is a polynomial of degree $d(\geq 2)$. Then $f'(z) - a = g'(P(z))P'(z)$. From Eq. (4.66) g' has an infinite number of zero $\{w_n\}$. For sufficiently large n, the equation $P(z) = w_n$ has d distinct roots that are also common roots of the simultaneous equations

$$\begin{cases} f(z) - az = g(w_n) \\ f'(z) - a = 0 . \end{cases}$$

This contradicts the conclusion of Lemma 4.8. This also shows that $f(z) - az$ is prime in enitre sense for every $a \in \mathbb{C} \backslash E$.

We now prove that for every $a \in \mathbb{C} \backslash E_1$ with at most one exception, $f(z) - az$ is prime. This is sufficient to show that $f(z) - az$ is not periodic for every $a(\in \mathbb{C})$ with at most one exception. If there are two distinct complex numbers a and b, such that both $f(z) - az$ and $f(z) - bz$ are periodic with periods τ and μ, respectively; then $f'(z)$ would have periods τ and μ. Hence τ/μ must be a real number. Thus $f(z) - az$ and $f(z) - bz$ both would be bounded on the straight line $\{t\tau, t \in (-\infty, \infty)\}$. Clearly, this is impossible. Theorem 4.9 is thus proved.

Theorem 4.10. (Noda) Let $f(z)$ be a transcendental entire function. Then the set

$$\{a | a \in \mathbb{C}; \ (z - a)f(z) \text{ is not prime}\}$$

is at most a countable set.

We need the following lemma.

Lemma 4.9. Let $f(z)$ be a transcendental entire function. Then there exists at most one countable set E such that for any nonzero constant $c(\in \mathbb{C})$ and $a \in \mathbb{C} \backslash E$ the simultaneous equations

$$\begin{cases} (z - a)f(z) = c \\ \dfrac{d}{dz}[(z - a)f(z)] = 0 \end{cases}$$

have at most one common root.

Proof. We unite

$$h(z) = z + f(z)/f'(z)$$

and

$$A = \mathbb{C} \backslash \{z \in \mathbb{C} | z \text{ is a zero or a pole of } h'(z)\} \ .$$

One can choose an open covering $\{C_i\}$ of A that satisfies the following three conditions:

(i) $\bigcup_i C_i = A$,
(ii) $h(z)$ is univalent in $C_i (i = 1, 2, \dots)$,
(iii) $D_i = \{h(z) | z \in C_i\}$ is a disk $(i = 1, 2, \dots)$.

Let

$$H(z) = (z - h(z))f(z) \ , \tag{4.67}$$
$$U_i(w) = (h/C_i)^{-1}(w) \ (w \in D_i, i = 1, 2, \dots);$$

namely

$$h \circ U_i(w) = w \ (w \in D_i; i = 1, 2, \dots) \ ; \tag{4.68}$$

$$V_i(w) = H(U_i(w)) \ (w \in D_i, i = 1, 2, \dots) \ , \tag{4.69}$$

$$I = \{(i, j), D_i \cap D_j \neq \phi, \ V_i(w) \not\equiv V_j(w), \ w \in D_i \cap D_j\} \ , \tag{4.70}$$

$$S_{ij} = \{w | w \in D_i \cap D_j; \ V_i(w) = V_j(w), \ (i, j) \in I\} \ , \tag{4.71}$$

and

$$\dot{E_0} = \left(\bigcup_i D_i\right)\backslash\left(\{h(z)|h'(z) = 0\} \cup \left\{\bigcup_{(i,j)\in Z} S_{i,j}\right\}\right.$$
$$\left.\cup\{z \in D_i|f(U_i(z)) = 0\}\right) .$$

Using arguments similar to those used in the previous lemma, we are going to derive the following four facts:

(a) $E = \mathbb{C}\backslash E_0$ is a countable set.

(b) $V_k(w) = (U_k(w) - w)f(U_k(w))$ $(w \in D_k)$

(c) If $V_i(a) = V_j(a)$ for some $a \in E_0$, then $U_i(a) = U_j(a)$.

(d) If $a \in E_0$, then the set $\{U_k(a); a \in D_k, k = 1, 2, \ldots\} \supseteq \{z|\frac{d}{dz}(\varsigma - a)f(\varsigma)|_{\varsigma=z} = 0$ but $(z - a)f(z) \neq 0\}$.

(a) and (b) are immediate consequences of Eqs. (4.67) through (4.71).

Next we shall show (c). From Eqs. (4.67) and (4.68) we deduce $V_i(w) = V_j(w)(w \in D_i \cap D_j)$. Thus $V_i'(a) = V_j'(a)$. From Eqs. (4.67), (4.68) and the fact that $H'(z) = -f(z)h'(z)$, we have

$$V_j'(a) = -f(U_i(a)) , \quad V_j'(a) = -f(U_j(a)) . \tag{4.72}$$

It follows from Eq. (4.71) that $f(U_i(a)) \neq 0 \neq f(U_j(a))$. This result and (b) yield $U_i(a) = U_j(a)$ and (c) is proven.

Now we prove (d). If $\frac{d}{d\varsigma}(\varsigma - a)f(\varsigma)|_{\varsigma=z_0} = f'(z_0)(z_0 - a) + f(z_0) = 0$ and $(z_0 - a)f(z_0) \neq 0$ for some $z_0 \in \mathbb{C}$, then $f'(z_0) \neq 0$. Consequently $a = z_0 + f(z_0)/f'(z_0) = h(z_0)$. From Eq. (4.71) we see that $h'(z_0) \neq 0$. Here $z_0 \in C_k$ for some positive integer k. This implies that $z_0 = U_k(a); a \in D_k$ and (d) is proven.

Proof of Theorem 4.10. Set

$$h(z) = z + \frac{f(z)}{f'(z)} ,$$

$$F_a(z) = (z - a)f(z) ,$$

and

$$E_1' = \{z|f(z) = 0\} \cup \{h(z)|h'(z) \neq 0\} .$$

Then for any $a \in \mathbb{C} \backslash E_1'$, $F_a(z)$ has at least one simple zero and

$$N(r, a, h) = \overline{N}(r, a, h) \le N(r, 0, F_a) - (N(r, 0, F_a) - \overline{N}(r, 0, F_a)) .$$

Let $t \in (0, \frac{1}{3})$. By the second fundamental theorem,

$$N(r, a, h) > tT(r, h) ,$$

holds for a set of r values of infinite measure for every complex number a with at most two exceptions. Further we see that for some positive k

$$T(r, h) \sim kT(r, F_a'/F_a) .$$

Now we recall a result of Gross, Osgood and Yang's. For transcendental entire function $F(z)$, $(z - a)F(a)$ cannot be factorized as $g(P(z))$, where P is a nonlinear polynomial and g is a transcendental entire function, for any complex number $a \in \mathbb{C} \backslash E$. Since this is so E is a countable set of complex numbers.

From this fact, with Lemma (4.9), and the remarks of Theorem 4.8, we deduce that there exists a countable set E_2' such that $F_a(z)$ is E-prime for every $a \in \mathbb{C} \backslash E_2'$. It is easy to verify that there exists at most a countable set E_3' such that $F_a(z)$ is not periodic for every $a \in \mathbb{C} \backslash E_3'$. Thus $F_a(z)$ is prime for every $A \in \mathbb{C}(E_2' \cup E_3')$. Theorem 4.10 is proved.

It does not seem to be a difficult task to find an example when an entire function $F(z)$ and $F(z) + P(z)$ are both not pseudo-prime for some non-constant polynomial. It is easy to find an entire function F and a polynomial $P(z)$ ($\not\equiv$ constant) so that $F(z)$ is pseudo-prime but $F(z) + P(z)$ is not.

When the order of F is restricted to be finite, the above observations may not be valid.

Conjecture 3. Let $F(z)$ be a transcendental entire function of finite order, that is pseudo-rpime. Then for any polynomial $P(z)(\not\equiv 0)$, $F(z) + P(z)$ and $P(z)F(z)$ remain pseudo-prime.

Research Problems. $f(z) = e^z$ is a finite-order, composite, entire function (i.e., not a prime function), but where $f(z) + z$ is prime. In addition $f(z) = e^{e^z}$ is a composite, entire function of infinite order, but $f(z) + z$ is prime. There naturally arises a question: Is there an entire function f, such that f and $f + z$ are both composite and not pseudo-prime? The answer is "yes". One can choose $f(z) = c^z + e^{e^z + z}$ we note that in the example f is of infinite order. We may further ask:

(i) Does there exist a finite order, entire function f such that f and $f + z$ are both composite but not pseudo-prime.

(ii) Keeping in mind the distribution of prime numbers in the set of integers, we may propose the following:

Let f be an entire function. For any two nonlinear polynomials p, q of different degrees, either $f + p$ or $f + q$ must be pseudo-prime. Furthermore, if p and q are relatively prime to each other, then one of the two $f + q$ and $f + p$, must be prime.

(iii) It is not difficult to show that if both f and g are entire and pseudo-prime, the product $f \cdot g$ may not be pseudo-prime.

(iv) If f is pseudo-prime, must $f(P)$ also be pseudo-prime for any polynomial $P(z)$?

(v) It has been shown by Song and Huang recently that to any pseudo-prime entire function f, f^n (n odd integer) always remains to be pseudo-prime. For even n, the above statement may not be true. If we choose $f(z) = \cos z e^{\sin z}$ as exhibited by Song-Huang, then f is prime (in entire sense), but $f^2 = \cos^2 z e^{2 \sin z} = (1 - w^2)e^{2w} \circ \sin z$.

(vi) Let $f(z)$ be a transcendental entire function and $\alpha(z)$ be an entire function ($\not\equiv 0$) satisfying $T(r, \alpha) = oT(r, f)$ as $r \to \infty$. Does it follow that $\alpha(z)f^2(z) + az$ is prime for any constant $a(\neq 0)$?

This would be a question more general than Gross' conjecture raised at the beginning of this chapter.

4.6. THE PSEUDO-PRIMENESS OF SOLUTIONS OF DIFFERENTIAL EQUATIONS

The prime function $e^z + P(z)$ (P a non-constant polynomial) or pseudo-prime function $\cos z$ and other similar forms of function all satisfy the following type of differential equation:

$$a_0(z)w^{(n)}(z) + a_1(z)w^{(n-1)}(z) + \ldots + a_{n-1}(z)w(z) + a_n(z) = 0 ,$$

where all the a_i ($i = 0, 1, 2, \ldots, n$) are polynomials.

In our previous discussions, we have employed fragmented and rather lengthy arguments to prove respectively the primeness of $e^z + P(z)$ and pseudo-primeness of $\cos z$. In 1980, N. Steinmetz obtained a general result and made the test of pseudo-primeness of many functions much easier.

Theorem 4.11. Let n be a positive integer and

$$w^{(n)}(z) + A_n(z)w^{(n-1)}(z) + \ldots + A_1(z)w(z) + A_0(z) = 0 , \qquad (4.73)$$

be a linear differential equation in w with all the coefficients $A_i(z)(i = 0, 1, 2, \ldots n)$ being rational functions. Then any meromorphic solution $h(z)$ of (4.73) must be pseudo-prime.

This theorem is a simple application of the main result in Steinmetz's 1980 paper, that will be stated and proven later. Let us briefly describe how Theorem 4.11 is proven.

First recall (in Wittich [24], p 73) that under the stated conditions of Eq. (4.73), any transcendental meromorphic solution of it has a positive and finite order. Now let $h(z)$ be such a solution and $h(z) = f(g(z))$; where f is meromorphic and g is entire. After substituting $h(z) = f(g(z))$ into Eq. (4.73) and combining terms we have

$$f^{(n)}(g)P_n(g) + f^{n-1}(g)P_{n-1}(g) + \ldots + f(g)P_0(g) + f_0(g)A(z) = 0 ,$$

where $P_i(g)(i = 1, 2, \ldots, n)$ denotes a differential polynomial in g with constants as the coefficients, $f_0(z) \equiv 1$, and $A(z)$ is a rational function. Theorem 4.12 (below), allows for the implication that $f^{(n)}, f^{(n-1)}, \ldots f$ satisfies an equation of the form Eq. (4.73) with the coefficients $A_i(z)$ being rational functions. Thus if f is not rational, then it can only be a transcendental meromorphic function of positive finite order. It follows from a well-known result of Edrei and Fuchs that g must be a polynomial. This proves Theorem 4.11.

Theorem 4.12 (Steinmetz) Let $F_0, F_1, \ldots, F_m$ be not identically vanishing entire functions and let $h_0, h_1, \ldots, h_m(m \geq 1)$ be arbitrary meromorphic functions not all identically zero. Let g be a non-constant entire function and satisfy the following condition:

$$\sum_{i=0}^{m} T(r, h_i) \leq kT(r, g) + S(r, g) , \qquad (4.74)$$

where k is a positive constant and $S(r, g)$ defines any quantity satisfying $S(r, g) = o(1)T(r, g), r \to \infty, r \notin E$; and where E is a set of r of finite measure (not necessarily the same at each occurrence). If F_i and h_i $(i = 0, 1, 2, \ldots, m)$ satisfy the following identity:

$$F_0(g)h_0 + F_1(g)h_1 + \ldots + F_m(g)h_m \equiv 0 , \tag{4.75}$$

then

(i) there exists polynomials in z, $P_0, P_1, \ldots, P_m$ not all zero such that

$$P_0 F_0 + P_1 F_1 + \ldots + P_m F_m \equiv 0 ; \tag{4.76}$$

(ii) there exists polynomials $Q_0, Q_1, \ldots, Q_m$ not all identically zero such that

$$F_0 Q_0 + F_1 Q_1 + \ldots + F_m Q_m \equiv 0 . \tag{4.77}$$

Proof. The proof consists of two main parts.* The first part is the construction of an auxiliary function for $s = 1, 2, \ldots$

$$H_s(z, t) = \frac{\sum_{i=0}^{m} P_{is}(g(z), t) F_i(t) h_i(z)}{Q_s(g(z))(g(z) - t)}$$

where $Q_s(\varsigma) = (\varsigma - t_1)(\varsigma - t_2) \ldots (\varsigma - t_{s-1})$ $(Q_1(\varsigma) \equiv 1)$ is a polynomial of degree s, t_i's are distinct constants not equal to t (t is a parameter), $P_{is}(\varsigma, t)$ is a polynomial in ς and t satisfies $P_{is}(t, t) \neq 0$; the degree of P_{is} denoted as δ_{is} in ς is independent of t. Moreover, whenever the denominator $(g(z) - t_1) \ldots (g(z) - t_{s-1})$ of $H_s(\varsigma, t)$ vanishes (i.e., $g(z) = t_k(1 \leq t \leq s-1)$), so does the numerator of the $H_s(\varsigma, t)$. This will be called the vanishing condition.

The second part is the proof that for s sufficiently large, the function $H_s(\varsigma, t)$ will vanish identically. The required polynomials $P_i(z)$ in (i) will be the corresponding polynomials $P_{is}(g(\varsigma), z) h_i(\varsigma)$ in $H_s(\varsigma, z)$.

We carry out the construction of the family $\{H_s(\varsigma, t)\}$ inductively and in such a fashion that the polynomial P_{is} has coefficients that are meromorphic functions of $t_1, t_2, \ldots, t_{s-1}$.

Moreover we can arrange $P_{is}(g, t)$ so that

$$\deg_g P_{is}(g, t) \text{ (the degree of } P_{is}(g, t) \text{ in } g) \leq s - 1 - \left[\frac{s-1}{m}\right] , \tag{4.78}$$

---

*The argument present here is essentially due to W.D. Brownawell which enables the result to be generalized to allow $g(z)$ to be a function of several complex variables.

where $[\alpha]$ denotes the largest integer $\leq \alpha$,

$$\deg_t P_{is}(g,t) \leq s - 1 \,, \tag{4.79}$$

and

$$P_{is}(t,t) = A_k(t) \text{ (the coefficients of } A_k(t) \text{ are} \tag{4.80}$$

meromorphic functions of $t_1, t_2, \dots, t_{s-1}$). Let

$$H_1(z,t) = \frac{F_1(t)h_1(z) + F_2(t)h_2(z) + \dots + F_m(t)h_m(z)}{g(z) - t} \,. \tag{4.81}$$

The parameter t will not assume the zeros or poles of F_i nor the roots of $g(z) - t$ that are poles of h_j for some j.

The vanishing condition is then satisfied by Eq. (4.75). Also the conditions stated in Eqs. (4.78), (4.79) and (4.80) hold trivially. We now define

$$H_2(z,t) = H_1(z,t_1) - a_1(t)H_1(z,t) \,,$$

where $a_1(t)$ is a suitably chosen meromorphic function of t_1 and t so that the conditions stated in Eqs. (4.78), (4.79) and (4.80) will be satisfied. In this way, we have constructed an infinite family of auxiliary functions $\{H_s(z,t)\}_{s=1}^{\infty}$. We have

$$H_1(z,t_1) = \frac{\sum_{i=0}^{m} F_i(t_1)h_i(z)}{(g(z) - t_1)} \,.$$

Thus,

$$
\begin{aligned}
&H_1(z,t_s) - a_1(t)H_1(z,t) \\
&= \frac{\sum_{i=0}^{m} F_i(t)h_i(z)}{g(z) - t_1} - \frac{\sum_{i=0}^{m} a_1(t)F_i(t)h_i(z)}{g(z) - t} \\
&= \frac{\sum_{i=0}^{m} \{F_i(t_1)(g(z) - t) - a_1(t)(g(z) - t_1)F_i(t)\}h_i(z)}{(g(z) - t_1)(g(z) - t)} \\
&= \frac{\sum_{i=0}^{m} P_{i2}(g(z),t)F_i(t)h_i(z)}{Q_2(g(z))(g(z) - t)} \,.
\end{aligned}
$$

In this case we can simply choose $a_1(t) = 1$. When $g(z) = t_1$, the numerator in the above function becomes

$$
\sum_{i=0}^{m} \{F_i(g(z))(t_1 - t) - (t_1 - t)F_i(t)\}h_i(z)
$$

$$
= \sum_{i=1}^{m} \{(t_1 - t)F_i(g(z))h_i(z)\}
$$

$$
= (t_1 - t)\sum_{i=0}^{m} F_i(g(z))h_i(z) \equiv 0 \,.
$$

The vanishing condition is thus satisfied. Also the conditions found in Eqs. (4.78), (4.79) and (4.80) hold trivially. We note from the above analysis that the larger the value of s, the more freedom will be obtained in deducing the degree of $P_{i_s}(g,t)$ in degree of g. We therefore see that one can inductively define

$$H_{s+1}(z,t) = H_1(z,t) - a_s(t)H_s(z,t) ,$$

where $a_s(t)$ is a meromorphic function in $t_1, t_2, \ldots, t_{s-1}$ and t so that the vanishing condition as well as the conditions in Eqs. (4.78), (4.79) and (4.80) will be satisfied.

Now let q be an integer so that

$$q > 2 + 2k , \tag{4.82}$$

where k is the constant defined in Eq. (4.74). We are going to show that whenever

$$s > (q+1)m + 1 , \tag{4.83}$$

$H_s(z,t) \equiv 0$. Now, suppose that this assertion is not valid. This means there exists at least an auxiliary function $H(z) \equiv H_s(z,t_s)$ with s satisfying Eq. (4.83) but $H(z) \not\equiv 0$. We write (while noting that $t_1, t_2, \ldots, t_3$ are constants),

$$H(z) = \frac{\sum_{i=0}^{m} P_i(g(z))h_i(z)}{Q(g(z))} .$$

The important fact is that according to Eq. (4.83) we have

$$\deg P_i(g) \leq s - 1 - \left[\frac{s-1}{m}\right] \leq s - 1 - \left[\frac{(q+1)m}{m}\right]$$
$$\leq (s-1) - (q+1) = s - q . \tag{4.84}$$

Now $Q(\varsigma) = (\varsigma - t_1)(\varsigma - t_2)\ldots(\varsigma - t_s)$ and is not identically zero. Let

$$A(\varsigma) = (\varsigma - t_1)(\varsigma - t_2)\ldots(\varsigma - t_q) ,$$
$$B(\varsigma) = (\varsigma - t_{q+1})(\varsigma - t_{q+2})\ldots(\varsigma - t_s) ,$$
$$F(z) = A(g(z)) ,$$

and

$$G(z) = H(z)F(z) .$$

By a standard argument we derive,

$$qT(r,g) = T(r, A(g)) + O(1) = T(r, F) + O(1)$$

$$= T\left(r, \frac{G}{H}\right) + O(1)$$

$$\leq T(r, G) + T(r, H) + O(1) . \tag{4.85}$$

By the constructions of the functions G and H and the vanishing condition we have

$$N(r, \infty, G) + N(r, \infty, H) \leq 2 \sum_{j=1}^{s} N_1^*\left(r, \frac{1}{g - t_j}\right) + 2 \sum_{i=0}^{m} N(r, \infty, h_i) \tag{4.86}$$

where $N_1^*(r, \infty, f)$ is the counting function for the poles of f with the multiplicities of poles of f being counted at least one less from their multiplicities. To estimate the proximity function, we note that since the degree of $P_i(g) \leq s - q$ $(i = 0, 1, 2, \ldots, m)$, the quotient

$$\left|\frac{P_i(\varsigma)}{B(\varsigma)}\right| \quad \text{and} \quad \left|\frac{P_i(\varsigma)}{Q(\varsigma)}\right|$$

are uniformly bounded for $|\varsigma - t_i| \geq 1$ $(i = 1, 2, \ldots, s)$. By simple calculations, we have

$$m(r, G) - m(r, H) \leq \sum_{j=1}^{s} m\left(r, \frac{1}{g - t_j}\right) + 2 \sum_{i=0}^{m} m(r, h_i) + O(1) . \tag{4.87}$$

It follows from Eqs. (4.85), (4.86) and (4.87) that

$$qT(r,g) \leq 2 \sum_{j=1}^{s} \left\{ m\left(r, \frac{1}{g - t_j}\right) + N_1^*\left(r, \frac{1}{g - t_j}\right) \right\} + 2 \sum_{i=0}^{m} T(r, h_i) + O(1) . \tag{4.88}$$

By the second fundamental theorem we have

$$\sum_{j=1}^{s} m\left(r, \frac{1}{g - t_j}\right) \leq 2T(r, g) - N_1(r, g) + S(r, g) \tag{4.89}$$

where

$$N_1(r, g) = N\left(r, \frac{1}{g'}\right) + 2N(r, g) - N(r, g') .$$

By definition, we have

$$2 \sum_{j=1}^{s} N_1^* \left(r, \frac{1}{g - t_j} \right) \leq 2N \left(r, \frac{1}{g'} \right) . \tag{4.90}$$

By combining Eqs. (4.74), (4.88), (4.89), and (4.90), and by noting that $N(r, g') - 2N(r, g)$ is negative, we obtain

$$qT(r, q) \leq 2T(r, g) + N(r, g') - 2N(r, g) + 2kT(r, g) + s(r, g)$$
$$\leq (2 + 2k)T(r, g) + s(r, g) . \tag{4.91}$$

This leads to

$$q \leq 2 + 2k ,$$

which contradicts Eq. (4.82), and proves Theorem 4.12.

Remarks. Gross and Osgood asserted that Theorem 4.12 remains valid if the conditions of Eq. (4.74) are replaced by the following two conditions:

(i) $\sum T(r, h_i) \leq kT(r, g)$,
on a sequence of r values $\{r_j\} \uparrow \infty$;

(ii) on the sequence $\{r_j\}$
$T(r, g') \leq (1 + o(1))T(r, g)$.

This remark will be used in the study of the relationship between the factorization and the existence of fix-points of an entire function of infinite order. (Theorem 4.15) and is useful in investigating problems on permutability of two entire functions.

Remark. Recently Gross-Osgood presented a simpler and transparent proof of Theorem 4.12 (see at the end of the appendix) by utilizing techniques used in studying theory of transcendental members and involving Nevanlinna's first fundamental theorem only.

Many elementary transcendental meromorphic functions $w = h(z)$ satisfy algebraic differential equations of the form:

$$\Omega(z, w, w', \dots, w^{(n)}) = 0 \tag{4.92}$$

where Ω is a polynomial in $w, w', \dots, w^{(n)}$ with rational functions as the coefficients. The linear differential equation is a special case of the Eq. (4.92). Let $M[h] = a(z)h^{k_0}(h')^{k_1} \dots (h^{(m)})^{k_m}$ with $a(z)$ a rational function $\not\equiv 0$

and is called a monomial in f. Then $\Omega(z, w, w', \ldots, w^{(n)})$ can be expressed as $\Omega[w] = \sum_{j=0}^{t} M_j[w]$ where $M_j[w]$ are monomials in f. Let $\gamma_m = k_0 + k_1 + \ldots + k_m$ and $\Gamma_m = k_0 + 2k_1 + \ldots + (m+1)k_m$. Define $\gamma_\Omega = \max_{j=0}^{t} \gamma_{M_j}$ and $\Gamma_\Omega = \max_{j=0}^{t} \Gamma_{M_j}$. It is not difficult to realize, based on Theorem 4.12, if a meromorphic function $h(z)$ satisfies algebraic differential of the form $\Omega(z, w, w', \ldots, w^{(n)}) = 0$, and if $h(z) = f(g(z))$, where f is transcendental meromorphic and g is entire; then f also satisfies an algebraic equation $\hat{\Omega}(z, w, w', \ldots, w^{(n)}) = 0$ of the form in Eq. (4.92) with $\gamma_{\hat{\Omega}} \leq \gamma_\Omega$ and $\Gamma_{\hat{\Omega}} \leq \Gamma_\Omega$. We refer the reader to Steinmetz's paper for more discussions on the pseudo-primeness of the solutions of certain special types of algebraic differential equations.

Now we raise some research problems on the pseudo-primeness of solutions of functional equations or difference equations.

(1) let $H(z) = \frac{1}{\Gamma(z)}$ th reverse function of the gamma function. It is well known that $H(z)$ is entire. It can be shown from the distribution of its zeros, that H is pseudo-prime. Moreover, $H(z)$ satisfies the following type of functional equations:

$$H(z + 1) = \frac{1}{z} H(z) \ .$$

We would like to know whether conditions similar to those of Theorem 4.11 can be derived for the above equation or a more general type of functional equations?

(2) Does there exist a finite order entire function f that is not pseudo-prime but satisfies a nonlinear differential equation with rational coefficients?

Song and Yang obtained some generalization of Theorem 4.11. The results will enable us to determine the pseudo-primeness of combinations of several meromorphic functions that are solutions of linear differential equations. We shall consider functions of the form

$$F(z) = \sum_{i=1}^{m} q_i(z) \psi_i(z) \ , \tag{4.93}$$

where $q_i(z)$ $(i = 1, 2, \ldots m)$ denotes a rational function and $\psi_i(z)$ a meromorphic solution satisfying the following type of linear differential equation:

$$w^{(n)}(z) + a_n(z) w^{(n-1)}(z) + \ldots + a_1(z) w(z) + a_0(z) = 0 \ , \tag{4.94}$$

where the $a_i(z) (i = 0, 1, 2, \ldots, n)$ are rational functions.

Let M be the family of all meromorphic functions and R the family of all rational functions. Clearly, M is a linear space over R, i.e., if $\varphi_1, \varphi_2 \in M$ and $R_1 \in R$, then $R_1\varphi_1 + \varphi_2 \in M$. Let D denote the class of meromorphic functions that satisfy a linear differential equation of the form as in Eq. (4.94). Also, for each $\varphi \in D$, we define the subset $L_\varphi = \langle 1, \varphi, \varphi', \varphi^{(n)} \rangle$, a set that is spanned by $1, \varphi, \varphi', \ldots, \varphi^{(n)}$ over R, when n is the minimal order of the different equation that is satisfied by φ.

It is easily verified that both D and L_φ are subspaces of M over R. Moreover we see that if $\varphi \in D$, then for any positive integer k, $\varphi^{(k)} \in D$. (In fact it can be shown that $\varphi^{(k)} \in L_\varphi \subset D$). Thus, any differential operator maps D into D.

Theorem 4.13. Let $F(z) = \sum_{j=1}^{m} q_j(z)\varphi_j(z)$, where $q_j \in R$ and $\varphi_j \in D; j = 1, 2, \ldots, m$.

Then $F(z)$ and all its derivatives $F^{(n)}, n = 0, 1, 2, \ldots$ are pseudo-prime.

Proof. Given that D is a subspace of M over R and $\varphi_i \in D$, we conclude that $F \in D$. Also as differential operators map D into D, this shows $F^{(k)} \in D(k = 0, 1, 2, \ldots)$. Hence by Theorem 4.11, $F^{(k)}$ is pseudo-prime for $k = 0, 1, 2, \ldots$.

If certain restrictions are placed on the growth rates of $\varphi_i (i = 1, 2, \ldots, m-1)$ and on the counting functions for the zeros and poles of $\varphi^{(m)}$, then we can prove that the function F stated in the above-theorem is actually left-prime.

Corollary 4.3. Suppose that in addition to the hypothesis of Theorem 4.13 the function ϕ_j satisfies

$$T(r, \varphi_j) = o(1)T(r, F) \text{ as } r \to \infty, j = 1, 2, \ldots, m-1 \quad (m \geq 2) \quad (4.95)$$

and

$$N(r, \varphi_m) + N\left(r, \frac{1}{\varphi_m}\right) = o(1)T(r, F) \text{ as } r \to \infty .$$

Then $F^{(k)} (k = 0, 1, 2, \ldots)$ is left-prime.

Proof. From Theorem 4.13 we know that $F^{(k)} (k = 0, 1, 2, \ldots)$ is pseudo-prime. We now show that F is left-prime. Suppose that $F = R_0(g)$, where g is transcendental entire, and $R_0(z)$ is a rational function of degree l. Therefore, we have $T(r, F) \sim lT(r, g)$. Set

$$\alpha(z) = \sum_{j=1}^{m-1} q_j(z)\varphi_j(z) .$$

By the hypothesis we obtain

$$T(r, \alpha) = o(1)T(r, F) = o(1)T(r, g) \text{ as } r \to \infty .$$

By Lemma 4.7 we have

$$\varlimsup_{\substack{r \to \infty \\ r \notin E}} \frac{N\left(r, \frac{1}{R_0(g) - \alpha}\right)}{T(r, g)} \geq l - 1 .$$

This will contradict with Eq. (4.95) unless $l - 1 = 0$, i.e., $R_0(\varsigma)$ has a bilinear form. This proves that F is left-prime.

Next we examine F'. Suppose that $F = R_1(h)$ with h being transcendental entire, and R_1 rational. We have

$$F'(z) = \left[\sum_{j=1}^{m} q_j \varphi_j(z)\right] = \sum_{j=1}^{m-1}(q_j' \varphi_j + q_j \varphi_j') + (q_m \varphi_m)' .$$

Set

$$\beta(z) = \sum_{j=1}^{m-1}(q_j' \varphi_j + q_j \varphi_j'), \quad \varphi(z) = q_m \varphi_m .$$

Then

$$F'(z) = \beta(z) + \psi'(z) .$$

In view of the proof for F, it is sufficient to show that

$$N(r, \psi') + N\left(r, \frac{1}{\psi'}\right) = o(1)T(r, g) . \tag{4.96}$$

Now,

$$N(r, \psi') \leq 2N(r, \varphi_m) = \circ T(r, F) = \circ T(r, g) ,$$

and

$$N\left(r,\frac{1}{\psi'}\right) + m\left(r,\frac{1}{\psi}\right) = N\left(r,\frac{1}{\psi'}\right) + m\left(r,\frac{\psi'}{\psi}\cdot\frac{1}{\psi'}\right) \qquad (4.97)$$

$$\leq N\left(r,\frac{1}{\psi'}\right) + m\left(r,\frac{1}{\psi'}\right) + m\left(r,\frac{\psi'}{\psi}\right) + O(1)$$

$$\leq T(r,\psi') + O(1) + m\left(r,\frac{\psi'}{\psi}\right)$$

$$\leq T(r,\psi) + N(r,\psi) + O(1) + m\left(r,\frac{\psi'}{\psi}\right)$$

$$\leq T\left(r,\frac{1}{\psi}\right) + N(r,\psi) + O(1) + m\left(r,\frac{\psi'}{\psi}\right)$$

$$\leq m\left(r,\frac{1}{\psi}\right) + N\left(r,\frac{1}{\psi}\right) + N(r,\psi) + m\left(r,\frac{\psi'}{\psi}\right) + O(1) .$$

Eliminating $m\left(r,\frac{1}{\psi}\right)$ from both sides of the above equations, we have

$$N\left(r,\frac{1}{\psi'}\right) \leq N\left(r,\frac{1}{\psi}\right) + N(r,\psi) + m\left(r,\frac{\psi'}{\psi}\right) + O(1) . \qquad (4.98)$$

We know from Wittich's result that every meromorphic solution φ_m of Eq. (4.94) is finite order. Hence $m\left(r,\frac{\psi'}{\psi}\right) = O(\log r)$.

Therefore Eqs. (4.95) and (4.98) yield Eq. (4.96). This proves that F' is left-prime. Using the above procedures, we can deduce immediately that $F^{(k)}\left(k = 1, 2, \ldots\right)$ are left-prime.

In summary, we have also proven that functions of the form $F(z) = \sin p(z) + q(z) J_n(z)$ where $p(z), q(z)$ are polynomials, $J_n(z)$ the nth Bessel function (namely $J_n(z)$ satisfies $w''(z) + \frac{1}{z}w'(z) + \left(1 - \frac{n^2}{z^2}\right) w(z) = 0$), is pseudo-prime. This result is not easily proved using the other prevailing arguments.

Several results concerning the pseudo-primeness of a function and its derivations obtained earlier by Ozawa, Yang, Urabe, and Niino now can be derived easily from Theorem 4.13.

For example, let

$$F(z) = \int_0^z (e^t - 1)e^{t^k} dt; \quad k \text{ an integer } \geq 3 ,$$

then $F^{(n)}\left(n = 0, 1, 2, \ldots\right)$ is E-prime.

Ozawa stated that it is possible to construct a family of prime functions using functions of the form

$$F(z) = \int_0^z Q(t)e^{P(t)}\,dt; \quad P, Q \text{ polynomials} .$$

Based upon the above discussions, the following general result can be derived.

Theorem 4.14. Let $P, P_1, P_2, \ldots, P_m$ and $Q_1, Q_2, \ldots, Q_m$ be polynomials with $Q_j \not\equiv 0, j = 1, 2, \ldots, m$ and deg $P >$ deg $P_m >$ deg $P_{m-1} > \ldots > \ldots >$ deg P_1.
Then the entire function

$$F(z) = \int_0^z \left[Q_1(t)e^{P_1(t)} + Q_2(t)e^{P_2(t)} + \ldots + Q_m(t)e^{P_m(t)} \right] dt + e^{P(t)}$$

is prime, unless there exist polynomials $\alpha, u, t'_j, s_j (j = 1, 2, \ldots, m)$ such that deg $\alpha \geq 2$ and

$$P = u(\alpha), \quad P_j = t_j(\alpha), \quad Q_j = s_j(\alpha); \quad j = 1, 2, \ldots, m .$$

Moreover, if $P(z)$ is prime, then F and all its derivatives $F^{(n)} (n = 1, 2, \ldots)$ are pseudo-prime.

The proof of this theorem is left to the reader.

Based on the above discussions, we pose the following conjecture:

Conjecture 4. Let $F(z)$ be a finite-order transcendental entire function. If F is E-pseudo-prime, then all its derivatives $F^{(n)} (n = 1, 2, \ldots)$ are E-pseudo-prime.

Question. What happens if F is of infinite order?

It is easy to construct an entire function F so that F and all its derivatives are composite (or all are prime).

Question. Does there exist an entire function $F(z)$ that possesses an infinite number of pseudo-prime functions as well as non pseudo-prime functions in the family $\{F^{(n)}(z); n = 0, 1, 2, \ldots\}$?

4.7. CONJECTURE 1 WITH $\rho(f(g)) = \infty$

In Sec. 4.2, it was proved that if f and g are nonlinear entire functions with $\rho(f \circ g) < \infty$ and at least one of them is transcendental, then $f(g(z))$

must have an infinite number of fix-points. Yang later extended this result so that f and g satisfy $\rho(f \circ g) = \infty$. This extension however, is subject to a variety of growth conditions. Thus Conjecture 1 has not been confirmed.[*] The most interesting result concerning Conjecture 1 that has been obtained is the following:

Theorem 4.15. (Gross and Osgood) Let f and g be two transcendental entire functions. Suppose that one of f and g is of finite order, while the other is of finite lower order. Then $f(g)$ must have an infinite number of fix-points.

Theorem 4.12 and its remarks will be applied in the proof of the theorem. Therefore, we shall first prove that, under the hypotheses, there exists a sequence of r values $\{r_n\}$ that will meet the requirements of the remarks. This is contained in the next lemma.

Lemma 4.10. Suppose that f and g are two transcendental entire functions. Suppose that f is of finite order of growth while g is of finite lower order, i.e., g satisfies, $\varliminf_{r \to \infty} T(r, g)/r^N = 0$ for some integer N. Suppose that $f(g)(z) = Q(z)e^{\alpha(z)} + z$, where Q is a polynomial and α is entire. Then there exists an unbounded monotone increasing sequence of r values $\{r_j\}_{j=1}^{\infty}$ such that
(i) for some constant $c, T(r_j, \alpha') < cT(r_j, g)$
and
(ii) in the sequence $\{r_j\}_{j=1}^{\infty}, T(r_j, g') = (1 + o(1))T(r, g)$.

Proof. From $\varliminf_{r \to \infty} T(r, g)/r^N = 0$, we may inductively construct a sequence of positive numbers $\{s_j\}_{j=1}^{\infty}$ such that for $j = 1, 2, \ldots, s_j > 8s_{j-1}$ and in $[1, s_j], T(r, g)/r^N$ assumes its minimal value at s_j.

From the well-known inequality

$$\log^+ (Mg(r/2)) \leq 3T(r, g) , \tag{4.99}$$

and Pólya's theorem that for any entire functions $h(z), k(z)$, there exists a positive constant $c(0 < c < 1)$ independent of r such that

$$M_h(cM_k(2r/3)) \leq M_{h(k)}(r) \leq M_h(M_k(r)) ,$$

we have

$$M_{e^z}(cM_\alpha(r/3)) < M_{e^\alpha}(r/2) . \tag{4.100}$$

Now from $Qe^\alpha + z = f(g)$ where f is a finite-ordered entire function, it is seen that there exist three positive numbers c_0, c_1 and c_2 independent of r such that

$$M_{e^\alpha}\left(\frac{r}{2}\right) \le r^{c_1} M_{f(g)}\left(\frac{r}{2}\right) \le \exp\left(c_2 Mg\left(\frac{r}{2}\right)\right)^{c_0} . \qquad (4.101)$$

It is seen that by applying log to both sides of Eqs. (4.100) and (4.101), for r sufficiently large we have,

$$c_3 \log\left(M_2\left(\frac{r}{3}\right)\right) \le c_4 \log M\left(Mg\left(\frac{r}{2}\right)\right) , \qquad (4.102)$$

where c_3, c_4 are two positive constants independent of r.

From this result and Eq. (4.99) we obtain

$$\log\left(M_\alpha\left(\frac{r}{3}\right)\right) \le c_5 T(r,g) , \qquad (4.103)$$

where c_5 is a positive constant.

By assuming that in $[1, s_j]$, $T(r,g)/r^N$ assumes its minimal value at s_j, thus, for $j = 0, 1, 2, \ldots$

$$T(s_j, g)/r_j^N \le T\left(\frac{s_j}{4}, g\right) / \left(\frac{s_j}{4}\right)^N$$

or

$$T(s_j, g) \le 4^N T\left(\frac{s_j}{4}, g\right) . \qquad (4.104)$$

From this result, coupled with Eqs. (4.103) and (4.104) we obtain:

$$\log\left(M_\alpha\left(\frac{s_j}{3}\right)\right) \le c_5 4^N T\left(\frac{s_j}{4}, g\right) . \qquad (4.105)$$

Since

$$M_{\alpha'}\left(\frac{r}{4}\right) \le O(1)\left(M_\alpha\left(\frac{r}{3}\right)\right) ,$$

and

$$T\left(\frac{r}{4}, \alpha'\right) \le \log\left(M_{\alpha'}\left(\frac{r}{4}\right)\right) ,$$

it follows from Eq. (4.105) that

$$T\left(\frac{s_j}{4}, \alpha'\right) \le O(1)T\left(\frac{s_j}{4}, g\right) . \qquad (4.106)$$

Hence, if we choose $r_j = \frac{s_j}{4}$, then $\{r_j\}$ satisfies (i).

Recall that for any entire function g and any positive $\rho > 1$

$$T(r, g') \leq T(r, g) + o(1)T(\rho r, g) . \qquad (4.107)$$

This and Eq. (4.104) imply that the sequence $\{r_j\}$ will also satisfy assertion (ii). This proves the lemma.

Proof of Theorem 4.15. Recall that if $f(g)$ has only a finite number of fix-points then $g(f)$ also has a finite number of fix-points. Therefore, we may assume without loss of generality that f is of finite order and g is of finite lower order.

Now assume $f(g)$ has only a finite number of fix-points. Thus

$$f(g(z)) - z = Q(z)e^{\alpha(z)} ,$$

where Q is a polynomial and α is an entire function ($\not\equiv$ constant). Differentiating both sides of the above equation we have

$$g' \cdot f'(g) - 1 = (Q' + \alpha'Q)e^{\alpha}$$
$$= \left(\frac{Q'}{Q} + \alpha'\right)(f(g) - z) .$$

Therefore

$$g' f'(g) - \left(\frac{Q'}{Q} + \alpha'\right) f(g) - 1 - z\left(\frac{Q'}{Q} + \alpha'\right) = 0 . \qquad (4.108)$$

Now it is easily verified that all the conditions of Theorem 4.12 and its remarks are satisfied for the above equation. By applying Theorem 4.12 to Eq. (4.108), where F_0, F_1 and F_2 are f', f and 1 respectively, we obtain:

$$P_1(z)f'(z) + P_2(z)f(z) + P_3(z) \equiv 0 , \qquad (4.109)$$

where P_1, P_2 and P_3 are relatively prime polynomials with $P_1 \cdot P_2 \cdot P_3 \not\equiv 0$. Substituting z by $g(z)$ in Eq. (4.109) yields

$$P_1(g)f'(g) + P_2(g)f(g) + P_3(g) = 0 . \qquad (4.110)$$

By eliminating $f'(g)$ from Eqs. (4.108) and (4.109) and letting $\gamma(z) = \frac{Q'}{Q} + \alpha'$, we obtain

$$P_1(g)(\gamma\{f(g) - z\} + 1) = g'\{P_2(g)f(g) + P_3(g)\} . \qquad (4.111)$$

Hence

$$[P_1(g)\gamma - g'P_2(g)]f(g) + P_1(g)(1 - z\gamma) - g'P_3(g) = 0 .$$

We claim

$$H(z) \equiv P_1(g)\gamma - g'P_2(g) \equiv 0$$

and

$$K(z) \equiv P_1(g)(1 - z\gamma) - g'P_3(g) \equiv 0 .$$

For otherwise, by solving Eq. (4.111) for $f(g)$, we have

$$f(g) \equiv \frac{-H}{K} \equiv \frac{-[P_1(g)(1 - z\gamma) - g'P_3(g)]}{P_1(g)\gamma - g'P_2(g)} ,$$

which implies

$$T(r, f(g)) = O(1)T(r, g) \text{ as } r \to \infty \text{ on the sequence } \{r_j\} .$$

This is impossible since f and g both are transcendental. Hence $H(z) \equiv 0 \equiv K(z)$. This yields

$$\{zP_2(g) + P_3(g)\}g' = P_1(g) ,$$

or

$$g' = \frac{P_1(g)}{zP_2(g) + P_3(g)} .$$

According to Steinmetz, Gackstatter and Laine's result (see the appendix), this implies that $zP_2(g) + P_3(g)$ is zero degree in g. Therefore both $P_2(g)$ and $P_3(g)$ are constants. Thus from Eq. (4.109) we have

$$P_1(z)f'(z) + \delta_1 f(z) + \delta_2 = 0 ,$$

where δ_1 and δ_2 are constants. It follows that $F - \frac{\delta_2}{\delta_1}$ can have a finite number of zeros. Hence $\delta\left(f(g), -\frac{\delta_2}{\delta_1}\right) = 1$. Therefore $\delta(f(g), z) = 0$, which implies that $f(g)$ must have an infinite number of fix-points. This concludes the proof of Theorem 4.15.

4.8. COMMON RIGHT FACTORS OF F AND $F^{(n)}$

Suppose that a meromorphic function $F(z)$ and its derivative $F'(z)$ have a transcendental entire function g as their common right factor, it is easily deduced from $F = f_1(g)$ and $F' = f(g)$ that g must be of the form $c_1 e^{c_2 z} + c_3$, where $c_i (i = 1, 2, 3)$ are constants. However, it is no longer a simple problem of searching for the possible forms of any common right factor of F and $F^{(n)} (n \geq 2)$. Earlier, Gross and Yang obtained the possible forms of common right factors for F, F'' and $F^{(iv)}$. Now as a result of Theorem 4.12, we can derive the following general result:

Theorem 4.16. (Steinmetz and Yang) Let $F(z)$ be a transcendental meromorphic function. Assume that $F = f(g)$ and $F^{(n)} = h(g)$; where $n \geq 2$, g a transcendental entire function and f and h are nonlinear meromorphic functions. Then (i) when $n = 2$, either g has the form $g(z) = c_1 e^{c_2 z} + c_3$ or $g(z) = c_1 \cos(c_2 z + c_3) + c_4$; where all the c_i are constants; or (ii) when $n > 2$, either f satisfies the following equation:

$$A_n(z) f^{(n)} + A_{n-1}(z) f^{(n-1)}(z) + \ldots + A_j(z) f^{(j)}(z) = 0 ,$$

where the $A_k(z) (j \leq k \leq n)$ are rational functions or g must assume one of the forms mentioned in (i).

Proof. We will first prove the case where $n = 2$. It follows from the assumptions $F = f(g)$ and $F'' = h(g)$ that we have

$$f''(g) g'^2 + f'(g) g'' = h(g) . \tag{4.112}$$

Application of Theorem 4.12 gives the following identity:

$$A(g) g'^2 + B(g) g'' + C(g) = 0 , \tag{4.113}$$

where $A(z)$, $B(z)$ and $C(z)$ are polynomials with $ABC \not\equiv 0$. Eliminating g'^2 from Eqs. (4.112) and (4.113), we get

$$[A(g) f'(g) - B(g) + f''(g)] g'' = A(g) h(g) + f''(g) C(g) . \tag{4.114}$$

If $A(g) f'(g) - B(g) f''(g) \not\equiv 0$, then from the above identity,

$$\begin{aligned} g'' &= [A(g) h(g) + f''(g) C(g)] / [A(g) f'(g) - B(g) f''(g)] \\ &= H_1(g) , \end{aligned}$$

where $H_1(z)$ is a meromorphic function. Clearly, H_1 cannot be transcendental. Furthermore, it is easily shown that H_1 must be a linear function. Hence we have

$$g'' = ag + b . \tag{4.115}$$

Substituting this equation into Eq. (4.112), we get

$$g'^2 = [h(g) - f'(g)(ag+b)]/f''(g) = H_2(g) .$$

Similarly, we can conclude that $H_2(z)$ must be a polynomial of degree ≤ 2. Thus

$$\begin{aligned} g'^2 &= t_1 g^2 + t_2 g + t_3 \\ &= t_1(g - s_1)(g - s_2) , \end{aligned}$$

where t_i and s_j are constants. It follows, depending on $s_1 = s_2$ or $s_1 \neq s_2$, that g assumes one of the forms stated in (i).

Now we consider the case: $A(g)f'(g) - B(g)f''(g) \equiv 0$, i.e.,

$$\frac{f''(w)}{f'(w)} = \frac{A(w)}{B(w)} . \tag{4.116}$$

Two cases will be considered separately: case (a): $A(w)$ is a constant and case (b): $A(n)$ is not a constant. We treat case (a) first. In this case, we may assume without loss of generality that $A(w) \equiv 1$. Equation (4.113) becomes

$$w'^2 B(w)w' + C(w) = 0 . \tag{4.117}$$

Set

$$B(w) = bw^{d_1} + B_1(w), \quad C(w) = cw^{d_2} + C_1(w) ,$$

where d_1, d_2 are the degrees of $B(w)$ and $C(w)$ respectively. Then, by Wittich's result on the existence theorem of solutions of certain differential equations, either $d_1 + 1 = d_2 > 2$ or $\max(d_1 + 1, d_2) = 2$. Suppose that $d_1 + 1 = d_2 > 2$, then by rewriting Eq. (4.117) as

$$w^{d_1}(bw'' + cw) = -w'^2 + B_1(w)w'' - C_1(w) \tag{4.118}$$

and applying Clunie's result [p. 68], we have

$$T(r, bw'' + cw) = m(r, bw'' + cw) = s(r, w) . \tag{4.119}$$

The central index $v(r)$ of g satisfies $b(v/z)^2 + c(1 + k_1(z)) = k_2(z)$, where $k_1(z)$ and $k_2(z)$ tend to zero as $|z| \to \infty$ (outside possibly a set of

$r(= |z|)$ values of finite length. Therefore the order of g is no greater than 1. However, according to a result of Ngoom and Ostrovskii's, we have

$$\varlimsup_{r \to \infty} \frac{m, \left(r, \frac{f'}{f}\right)}{\log r} = \max(t - 1, 0)$$

for any meromorphic function f of order t $(< \infty)$. Thus the term $S(r, g)$ in the equation has a magnitude $o(1) \log r$ and $bg'' + cg$ can only be a constant. This leads to the situation seen in Eq. (4.115) we encountered before. We now treat the situation: $\max(d_1 + 1, d_2) = 2$. Eq. (4.117) then becomes

$$w'^2 + (b_0 + b_1 w)w'' + c_0 + c_1 w + c_2 w^2 \equiv 0 , \tag{4.120}$$

where b_1 and c_1 are the leading coefficient of $B_1(w)$ and $C_1(w)$ respectively. Again by the central index theorem, we derive

$$(v/z)^2 + h_1(1 + h_1(z))(v/z) = h_2(z) , \tag{4.121}$$

where $h_1(z)$ and $h_2(z)$ tend to zero as $|z| \to \infty$, outside a set of $r(= |z|)$ values of finite length. In the meantime, we have

$$\frac{f''(w)}{f'(w)} = \frac{A(w)}{B(w)} = \frac{1}{b, w + b_0} .$$

It follows from examining the residue that $\frac{1}{b_1} \neq 0, -1$ and 1. We conclude from Eq. (4.121) that the order of g is ≤ 1. If g' never vanishes, then we are done. If we assume that $g'(z_0) = 0$ for some z_0, then be differentiating

$$w'^2 + B(w)w'' + C(w) = 0 .$$

By setting $z = z_0$, we get,

$$B(w_0)g'''(z_0) = 0; \quad w_0 = g(z_0) .$$

Two cases may arise: (i) $B(w_0) = 0$ and case (ii) $B(w_0) \neq 0$. If $B(w_0) = 0$, then $B(w) = b_0 w$. Substituting this result into Eq. (4.11) and letting $z = z_0$ yields $c_0 = 0$, and Eq. (4.120) becomes

$$w'^2 + b_1 w w'' + w(c_1 + c_2 w) = 0 . \tag{4.122}$$

Two subcases will be considered (ia) $c_1 = 0$ and (ib) $c_1 \neq 0$. Under (ia) by substituting $y = w'/w$ into Eqs. (4.122) we have,

$$b_1 y^2 + (1 + b_1) y^2 + c_2 = 0 . \qquad (4.123)$$

We note now that y has a simple pole at $z = z_0$ with residue p; where p is an integer ≥ 2. Comparing the coefficients of the term $\frac{1}{z - z_0}$ in the above equation, we see

$$-p b_1 + (1 + b_1) p^2 = 0 .$$

On the other hand, $\frac{1}{b_1} = q$ is also an integer $\neq 0, -1$, and 1. Then the above equation yields $(q + 1)p = 1$, which is a contradiction.

We now consider case (ib): Since the order of g is no greater than one, we are done if g never vanishes. So we assume $g(z_1) = 0$ for some z_1. Then it follows from Eq. (4.122) that $g'(z_1) = 0$, but $g''(z_1) \neq 0$ (since $c_1 \neq 0$). Thus every zero of g is of multiplicity 2. Hence $g(z) = K^2(z)$ for some entire function K and Eq. (4.12) becomes

$$b_1 K K'' + (b_1 + 2) K'^2 + \frac{c_1}{2} + \frac{c_2}{2} K^2 = 0 . \qquad (4.124)$$

Differentiating above equation we get

$$(3 b_1 + 4) K'(z_1) K''(z_1) = 0 .$$

Since $3b_1 + 4 \neq 0$ (as $\frac{1}{b_1}$ is an integer) and $K_1'(z_1) \neq 0$ we conclude $K_1''(z_0) = 0$. Therefore K''/K is an entire function (since $K = 0$ has only simple zeros). Thus $m\left(r, \frac{K''}{K}\right) = T\left(r, \frac{K''}{K}\right) = o(1) \log r$. It follows that $\frac{K''}{K}$ is a constant. Substituting this into Eq. (4.124) we get

$$K'^2 = d_1 K^2 + d_2, \quad d_1, d_2 \text{ constants} .$$

From $g = K^2$ we have $g' = 2K K'$ and, hence, $g'^2 = 4 K^2 K'^2 = 4g(d_1 g + d_2)$. This goes back to Eq. (4.115). Thus case (i) is settled completely. Now we discuss case (ii): $B(w_0) \neq 0$. Again if g' never vanishes then we are done. Therefore, we assume that $g'(z_0) = 0$ for some z. Then from Eq. (4.117) we can derive the same conclusion $g''(z_0) = 0$ $\left(\text{but } g''(z_0) \neq 0 \text{ by the}\right.$ uniqueness theorem for the equation $w'' = -\frac{C(w)}{B(w)} - \frac{w'^2}{B(w)}\right)$.

In a similar manner, we find that g''/g is entire and, moreover, it must be a constant. This leads to the form found in Eq. (4.115), that has been settled already. All the above discussions conclude the case where $A(w)$ is a constant. To complete the proof for the case $n = 2$ we need to settle case (b); that $A(w)$ is not a constant. We may assume that $A(0) = 0$ and shall treat two subcases separately: subcase (b1) $B(0) = 0$ and subcase (b2) $B(0) \neq 0$. Suppose that case (b1) holds. Then it may also be assumed, without loss of generality, that A, B, C are relatively prime. It follows, from $A(0) = B(0) = 0$, that $C(0) \neq 0$. we recall a result of A.Z. Mokhouko and V.D. Mokhouko. Suppose that $P(z, w, \dots, w^{(n)})$ is a differential polynomial in w with polynomials as the coefficients and that f is a transcendental, meromorphic function solution of $P(z, w, w', \dots, w^{(n)}) = 0$ with $P(z, 0, 0, 0, \dots, 0) \not\equiv 0$. Then

$$m\left(r, \frac{1}{f}\right) = S(r, f) .$$

Therefore, by applying this result to $P(z, w, \dots, w^{(n)}) \equiv A(w) + B(w)w'' + C(w)$ (and noting that $P(z, 0, 0, \dots, 0) = C(0) \not\equiv 0$), we get

$$m\left(r, \frac{1}{g}\right) = S(r, g) . \tag{4.125}$$

On the other hand; from $C(0) \neq 0, A(0) = B(0) = 0$ it becomes clear that g never vanishes. This contradicts Eq. (4.125). Then case (b1) has to be excluded. We now proceed to settle case (b2): Let z_0 be a zero of $g(z)$ with multiplicity t and $g''(z_0) = -C(0)/B(0) = d$ (a constant).

Then if $t = 1$,

$$K(z) = \frac{g''(z) - d}{g(z)} \left(\equiv \frac{w'' - d}{w}\right) , \tag{4.126}$$

will be regular at $z = z_0$. We are going to show that it is impossible to have $t \geq 2$. Otherwise, from Eq. (4.115), we can successively derive $g'(z_0) = g^n(z_0) = \dots = g^{(n)}(z_0) = 0, \forall n$. But $g(z) \not\equiv 0$ and we conclude that g has no multiple roots, and therefore $K(z)$ is an entire function. We show in fact that K must be a constant. Assume that K is not a constant. It then follows from Eqs. (4.126) and (4.125) that

$$T(r, K) = m(r, K) \leq m\left(r, \frac{g''}{g}\right) + m\left(r, \frac{d}{g}\right) .$$
$$= S(r, g) \tag{4.127}$$

Now rewrite the differential equation of g in Eq. (4.113) as

$$A(w)w'^2 + dB(w) + C(w) + K(z)wC(w) = 0 . \qquad (4.128)$$

If $K(z)$ is not a constant, then from Steinmetz's results ["Über die fakoriserbaren Lösungen gewöhnlicher differentialgleichungen", *Math. Zeit,* **170** (1980), Theorem 3], we can conclude that $A(w)$ is a common factor (in the product sense for $dB(w) + C(w)$ and $wC(w)$, that is,

$$wC(w) = A(w)A_1(w), \quad dB(w) + C(w) = A(w)A_2(w) ,$$

where A_1, A_2 are polynomials. Therefore Eq. (4.128) can be simplified as

$$w'^2 + A_2(w) + K(z)A_1(w) = 0 . \qquad (4.129)$$

Again, by using Steinmetz's result mentioned above, we get

$$A_1(w) = \alpha_0 + \alpha_1 w + \alpha_2 w^2 ,$$

and

$$A_2(w) = \beta_0 + \beta_1 w + \beta_2 w^2 ,$$

where α_i $(i = 0, 1, 2)$ and β_i $(i = 0, 1, 2)$ are constants. Differentiating Eq. (4.129) and substituting w'' by $d + Kw$, we obtain

$$[2d + b_1 + a_1 K + w(2K + 2b_2 + 2b_2 K)]w' = -K'[a_0 + a_1 w + a_2 w^2] .$$

Accordingly, we have

$$w' = K_1 + K_2 w , \qquad (4.130)$$

where K_1 and K_2 are meromorphic functions satisfying $T(r, K_i) = S(r, g)$, $i = 1, 2$. Replacing w'^2 in Eq. (4.129) by $K_1^2 + 2K_1 K_2 w + K_2^2 w^2$ then by comparing coefficients, we get

$$K^2 + \beta_2 + \alpha_2 K = 0 . \qquad (4.131)$$

On the other hand, by differentiating both sides of Eq. (4.130) and substituting w'' by $d + w', w' = K_1 + K_2 w$, we get

$$\begin{aligned}
d + Kw &= K_1' + K_2' w + K_2 w' \\
&= K_1' + K_2' w + K_2(K_1 + K_2 w)
\end{aligned} \qquad (4.132)$$

Comparing the constant term and the coefficients of the term w, we get

$$d = K_1' + K_2 K_1 , \qquad (4.133)$$

and

$$K = K_2' + K_2^2 . \qquad (4.134)$$

It follows from Eq. (4.131) that

$$T(r, K) = 2T(r, K_2) + O(1) .$$

But Eqs. (4.131) and (4.132) yield

$$K = K_2' + \beta_2 + \alpha_2 K . \qquad (4.135)$$

Hence

$$(1 - \alpha_2)K = K_2' + \beta_2 ,$$

and

$$T(r, K) = T(r, K_2') \le (1 + o(1))T(r, K_2) + S(r, K_2) . \qquad (4.136)$$

This contradicts Eq. (4.135) unless $1 - \alpha_2 = 0$, and therefore K_2' is a constant. It follows from Eq. (4.126) that K is also a constant. Therefore, Eq. (4.126) reduces to Eq. (4.115). This completes the proof of the part (i).

Now we prove part (ii); $n > 3$. Let

$$F(z) = f(g(z)) \text{ and } F^{(n)}(z) = h(g(z)) . \qquad (4.137)$$

Then

$$f^{(n)}(g)D_n[g] + f^{(n-1)}(g)D_{n-1}[g] + \ldots + f'(g)D_1[g] = h(g) , \qquad (4.138)$$

where $D_n[g] = g'^n, D_{n-1}[g] = \frac{n(n-1)}{2}(g')^{n-2}g'', \ldots, D_1[g] = g^{(n)}$. In general, $D_j[g]$ defines a homogeneous differential polynomial in g of degree j. Thus

$$T(r, D_j[g]) = m(r, D_j[g]) \le jm(r, g) + S(r, g) .$$

We may assume without loss of generality that $f^{(n)}(z) \not\equiv 0$. It is easily verified that Theorem 4.12 applies to Eq. (4.138). We have

$$D_n[g] = R_{1,n-1}[g]D_{n-1}[g] + \ldots + R_{1,1}(g)D_1[g] + R_{1,0}(g) , \qquad (4.139)$$

where the R_{ij} are rational functions. Equations (4.138) and (4.139) yield

$$\sum_{j=1}^{n-1} [f^{(j)}(g) + R_{1,j}(g)f^{(n)}(g)]D_j[g] = h(g) - f^{(n)}(g)R_{1,0}(g) . \qquad (4.140)$$

We apply Theorem 4.12 to the above equation and obtain:

$$D_{n-1}[g] = R_{2,n-2}(g)D_{n-2}[g] + \ldots + R_{2,1}(g)D_1[g] + R_{2,0}(g) . \qquad (4.141)$$

Similarly we derive from Eqs. (4.141) and (4.140)

$$\sum_{j=1}^{n-2} [f^{(j)}(g) + (R_{ij}(g) + R_{1,n-1}(g)R_{2,j}(g))f^{(n)}(g) + R_{2,j}(g)f^{(n-1)}(g)]D_j[g]$$

$$= h(g) - f^{(n)}(g)R_{1,0}(g) - f^{(n-2)}(g)R_{2,0}(g) . \qquad (4.142)$$

If the procedure does not break down, we finally arrive at

$$(g')^n = D_n[g] = R(g) , \qquad (4.143)$$

where R is a rational function. The procedure can only break down if at some stage Theorem 4.12 is not applicable. This implies that one of the functions occurring in Eqs. (4.140) or (4.142) vanishes identically, i.e.,

$$A_n(w)f^{(n)}(w) + A_{n-1}(w)f^{(n-1)}(w) + \ldots + A_j(w)f^{(j)}(w) \equiv 0 .$$

To complete the proof of Theorem 4.16 we have to show that Eq. (4.143) has no other solutions than that stated in part (i). It is easy to see that if Eqs. (4.143) holds then R must be, in fact, a polynomial, say;

$$R(w) = c(w - t_1)^{p_1}(w - t_2)^{p_2} \ldots (w - t_k)^{p_k} ,$$

where c, p_i and t_j are constants; p_i and k are positive integers. Consideration of the multiplicities of the zeros on both sides of Eq. (4.143) leads to the only two possible cases (i) $k = 1$ and $p_k = n$, (ii) $k = 2$ and $p_1 = p_2 = \frac{n}{2}$. It is easily seen that only solutions of (4.143) that leads to an entire g, is $g = c_1 e^{c_2 z} + c_3$ or $g = c_1 \cos(c_2 z + c_3) + c_4$. This also completes the proof of the theorem.

APPENDIX

1. FUNCTIONAL IDENTITIES

In a manner similar to that of the proof of Theorem 1.9, Ozawa and Hiromi were able to obtain the following result that has been applied to resolve many types of value-distribution problems.

Theorem A.1. Let $a_0(z), a_1(z), \ldots, a_n(z)$ be nonzero meromorphic functions and $g_1, g_2, \ldots, g_n$ be entire functions. Suppose that

$$T(r, a_j(z)) = o(1) \left(\sum_{i=1}^{n} T(r, e^{g_i}) \right), \quad j = 0, 1, 2, \ldots, n$$

and identity:

$$\sum_{i=1}^{n} a_i(z) e^{g_i(z)} \equiv a_0(z)$$

holds.

Then there must exist a set of constants $\{c_i\}_{i=1}^{n}$ not all zero, such that

$$\sum_{i=1}^{n} c_i a_i(z) e^{g_i(z)} \equiv 0 .$$

SOME SIMILAR RESULTS

(i) (Niino and Ozawa) Let $G_j(z)(j = 1, 2, \ldots, n)$ be transcendental entire functions, c_j $(j = 1, 2, \ldots, n)$ be nonzero constants, and $g(z)$ be an entire function. Suppose that

$$\sum_{j=1}^{n} c_j G_j(z) \equiv g(z) .$$

Then $\sum_{j=1}^{n} \delta(0, G_j) \leq n - 1$.

(ii) (Yang) Let f, g be transcendental entire functions and m, n be two positive integers ≥ 2. Let $a_1(z), a_2(z)$ be two meromorphic functions satisfying $T(r, a_i(z)) = o(1)T(r, f)$ and $T(r, a_i(z)) = o(1)T(r, g)$ as $r \to \infty, i = 1, 2$. Find a necessary condition for m, n so that the following identity holds

$$a_1 f^m(z) + a_2(z) g^n(z) \equiv 1 .$$

(iii) Toda, generalized the above problem and obtained the following result:

Let $f_0, f_1, \dots, f_p(p \geq 1)$ be $p + 1$ non-constant entire functions and $a_0(z), a_1(z), \dots, a_p(z)$ be $p + 1$ nonzero meromorphic functions satisfying $T(r, a_i(z)) = o(1)T(r, f_i)$ as $r \to \infty, i = 0, 1, 2, \dots, \infty$. If, for some non-negative integers $n_0, n_1, \dots, n_{p_1}$, the following identity holds:

$$\sum_{i=0}^{p} a_i(z) f_i^{n_i}(z) \equiv 1 ;$$

then it is necessary that

$$\sum_{i=0}^{p} \frac{1}{n_i} \geq \frac{1}{p} .$$

2. RELATIONS BETWEEN THE GROWTH OF $f(g), f$ AND g

One of the basic factorization problems is that of comparing the growths of $f(g), f$ and g. Among the many important relationships the following two are essential ratios:

$$A(R, f, g) = \frac{\log M(r, f(g))}{\log M(r, f) + \log M(r, g)} \tag{1}$$

for entire functions f and g; and

$$B(r; f, g) = \frac{T(r, f(g))}{T(r, f) + T(r, g)} \tag{2}$$

for f meromorphic and g entire.

For the sake of simplicity, we assume that all the functions f, g that appear in this section are transcendental unless stated otherwise. The corresponding results for the cases where one of the functions (f, g) are polynomial or a rational function will be quite clear.

Generally speaking, when f is entire, $\lim_{r\to\infty} B(r; f, g) = \infty$ and $\overline{\lim}_{r\to\infty}$ $A(r; f, g) = \infty$. For most meromorphic function f, $\lim_{r\to\infty} B(r; f, g) = \infty$ remains valid. But Clunie exhibited an example that shows $\lim_{r\to\infty} B(r; f, g) = \infty$ is not always true. In this section we shall be concerned with the behaviors (as $r \to \infty$) of the ratios

$$\log M(r, f \circ g)/\log M(r, f) \quad \text{and} \quad \log M(r, f \circ g)/\log M(r, g),$$

when $f(z)$ and $g(z)$ are both entire, and

$$T(r, f \circ g)/T(r, f), \quad \text{and} \quad T(r, f \circ g)/T(r, g)$$

when f is meromorphic and g is entire.

Before proceeding with the discussions of the above cases we introduce some preliminary results from Clunie that will be used later.

Theorem A.2. (Clunie) Let $f(z)$ and $g(z)$ be entire with $g(0) = 0$. Let ρ satisfy $0 < \rho < 1$ and let $c(\rho) = (1 - \rho)^2/4\rho$. Then for $R \geq 0$,

$$M(R, f \circ g) \geq M(c\rho M(\rho R, g), f) . \tag{3}$$

Proof. Let $R > 0$ be fixed. Then the complement of the set $\{w \mid g(z) = w; |z| = R\}$ in the W-plane is an open set and consists of a finite number of disjoint domains. Let D_R represent the unbounded one and let Γ_R define its boundary. It is easy to see that Γ_R is a Jordan curve. Let Δ_R define the bounded domain having Γ_R as its boundary. Since $g(0) = 0$, it follows for all small $\eta > 0$ that the circle $c_\eta = \{w \mid |w| = \eta\} \subseteq \Delta_R$. Let k be the largest number such that $|w| = k \subseteq \Delta_R \cup \Gamma_R$. Then

$$M(k, f) \leq \max_{w \in \Gamma_R} |f(w)| \leq M(R, f \circ g) \tag{4}$$

by the maximal modulus principle.

The theorem follows if $c(\rho)M(\rho R, g) \leq k$, and thus we now proceed to prove. Let γ_1 be a path that joins a point on $|w| = k$ to $w = \infty$ and lies apart from its end points in D_R. Now change the variables to $W = w/k$ and $\varsigma = z/R$ and let γ be the path in the W-plane corresponding to γ. Let $h(\varsigma)$ be a univalent conformal mapping from $|\varsigma| < 1$ onto the complement of γ and satisfy $h(0) = 0$. Then $g_1(\varsigma) = g(\varsigma R)/k$ maps $|\varsigma| < 1$ conformally onto $|W| < 1$ and $g_1(\varsigma)$ is subordinate to $h(\varsigma)$ in $|\varsigma| < 1$. This result implies that $\{g_1(\varsigma)\,|\,|\varsigma| < t\} \subseteq \{h(\varsigma)\,|\,|\varsigma| < t\} \forall\, 0 < t < 1$. Thus

$$\max_{|\varsigma|=\rho} \frac{|g(\varsigma R)|}{k} \leq \max_{|\varsigma|=\rho} |h(\varsigma)| \ . \tag{5}$$

Set

$$S(\varsigma) = h(\varsigma)/h'(0) \ .$$

Then S is univalent and analytic in $|\varsigma| < 1$ with $S'(0) = 1$. By Koebe's $\frac{1}{4}$ Theorem, the image of $|\varsigma| < 1$ by $S(\varsigma)$ covers the disk $|W| < \frac{1}{4}$ and $\max_{|\varsigma|=\rho} |S(\varsigma)| \leq \rho/(1-\rho)^2$. Thus the image of $|\varsigma| < 1$ by $h(\varsigma)$ covers the disk $|W| < |h'(0)|/4$. This image omits all the points on $|W| = 1$ and so $|h'(0)| \leq 4$. Consequently,

$$\max_{|\varsigma|=\rho} |h(\varsigma)| \leq \frac{|h'(0)|\rho}{(1-\rho)^2} \leq \frac{4\rho}{(1-\rho)^2} \ . \tag{6}$$

From this result and Eq. (5) it follows that

$$\frac{M(\rho R, g)}{k} \leq \frac{4\rho}{(1-\rho)^2} = \frac{1}{c(\rho)} \ , \tag{7}$$

and so $c(\rho)M(\rho R, g) \leq k$. This proves Theorem A.2.

A special and frequently used case of the above theorem is when $\rho = \frac{1}{2}$.

Corollary A.1. (Pólya) Let $f(z), g(z)$, and $h(z)$ be entire functions with $h(z) = f(g(z))$. If $g(0) = 0$, then there exists an absolute constant $c, 0 < c < 1$ such that for all $r > 0$ the following inequality holds:

$$M(r, h) \geq M\left(cM\left(\frac{r}{2}, g\right), f\right) \ . \tag{8}$$

Remarks. (i) The above inequality can also be expressed as

$$M_r(h) \geq M_f\left(cMg\left(\frac{r}{2}\right)\right) \ .$$

(ii) When $g(0) \neq 0$, the corresponding inequality should read,

$$M(r, f \circ g) \geq M\left(cM\left(\frac{r}{2}, g\right) - |g(0)|, f\right) \ .$$

(iii) c can be chosen to be $\frac{1}{8}$.

From the corollary we obtain the following important fact concerning the order and lower order of a composite function $f(g)$.

Theorem A.3. If $f(z)$ and $g(z)$ are two entire functions such that $f(g)$ is of finite order (lower order), then

(i) either $g(z)$ is a polynomial and $f(z)$ is of finite order (lower order),

or

(ii) $g(z)$ is not a polynomial but a function of finite order (lower order) and $f(z)$ is of zero order (lower order).

Proof. We shall prove the theorem for $f(g)$ of finite order and make a remark concerning the case when $f(g)$ is of finite lower order.

Clearly, we may assume that both f and g are non-constant and that $g(0) = 0$. From the hypotheses we have for any $\varepsilon > 0$

$$M_{f \circ g}(r) = O(1)e^{r^{\alpha+\varepsilon}} \, ,$$

where α is the order of f. Let

$$g(z) = \sum_{i=0}^{\infty} a_j z^j, \quad z = re^{i\theta}$$

be the Taylor expansion of g.

Then from

$$|g(z)|^2 = g(z)\overline{g(z)} = \sum_{n=0}^{\infty} |a_n|^2 r^{2n} \, ,$$

it follows that

$$M_g(r) \geq |a_m| r^m, \quad m = 0, 1, 2, \ldots \, .$$

By Theorem A.2 we have (for any $\varepsilon > 0$ and any $m > 0$).

$$M_f(c|a_m|2^{-m}r^m) \leq M_f\left(cMg\left(\frac{r}{2}\right)\right) \leq M_{f \circ g}(r) = O(e^{r^{\alpha+\varepsilon}}) \, .$$

The order of $f, \rho(f) \leq \alpha/m$. If g is not a polynomial ($m \to \infty$), we derive $\rho(f) = 0$. Furthermore, let

$$f(z) = \sum_{j=0}^{\infty} b_j z^j, \quad z = re^{i\theta} \, ,$$

then

$$M_f(r) \geq |b_n| r^n, \quad |b_n| > 0, \quad n \geq 1 \, ,$$

and

$$|b_n|c^n \left(M_g\left(\frac{r}{2}\right)\right)^n \leq M_f\left(cM_g\left(\frac{r}{2}\right)\right) \leq M_{f \circ g}(r) = O(e^{r^{\alpha+\epsilon}})$$

so that g is at most of finite order.

Remark. Pólya proved the theorem for $f(g)$ of finite order. Gross remarked that a similar argument will provide the result for $f(g)$ of finite lower order. Later Song and Yang noted it was not quite so. They gave a formal proof in the paper "Further Growth Properties of Composition of Entire and Meromorphic Functions" [*Indian J. Pure & Appl. Math.* **(15) 1** (Jan 1984) 67-82]. Later on Walter Bergweiler pointed out in his dissertion (1986) that Song-Yang's argument was incomplete and presented a different proof. More recently, he showed an interesting and related result: If f is meromorphic and g is entire, then

$$T(r, f \circ g) \leq (1 + o(1))\frac{T(r, g)}{\log M(r, g)}T(M(r, g), f) .$$

Theorem A.4. Let f and g be two transcendental entire functions. Then

$$\lim_{r \to \infty} \frac{\log M(r, f \circ g)}{\log M(r, f)} = \infty , \tag{9}$$

$$\lim_{r \to \infty} \frac{T(r, f \circ g)}{T(r, f)} = \infty \tag{10}$$

(when f is meromorphic, this result may not be valid),

$$\lim_{r \to \infty} \frac{\log M(r, f \circ g)}{\log M(r, g)} = \infty , \tag{11}$$

and

$$\lim_{r \to \infty} \frac{T(r, f \circ g)}{T(r, g)} = \infty . \tag{12}$$

(This result remains valid when f is meromorphic.)

Proof. We note that if $f(z)$ is entire and transcendental, then for r sufficiently large, $\log M(r, f)/\log r$ is an increasing function of r and a convex function of $\log r$. It follows that $\log M(r, f)/\log r \uparrow \infty (r \geq r_0)$. From Theroem A.2 and its remarks we have

$$
\begin{aligned}
\frac{\log M(r, f \circ g)}{\log M(r, f)} &\geq \frac{\log M \left(\frac{1}{8} M \left(\frac{r}{2}, g\right) - |g(0)|, f\right)}{\log M(r, f)} \\
&\geq \frac{\log \left(\frac{1}{8} M \left(\frac{r}{2}, g\right) - |g(0)|\right)}{\log r} \\
&= \frac{\log M \left(\frac{r}{2}, g\right)}{\log r} + o(1) \to \infty \quad \text{as } r \to \infty .
\end{aligned}
$$

This proves Eq. (9).

To prove Eq. (10), we recall the following well-known inequality:

$$
T(r, f) \leq \log M(r, f) \leq \frac{R + r}{R - r} T(r, f), \quad R > r \geq 0 ,
$$

for any non-constant entire function f. It follows that for large r

$$
T(r, f \circ g) \geq \frac{1}{3} \log M \left(\frac{r}{2}, f \circ g\right) .
$$

From the proof of Eq. (9), we have

$$
\frac{T(r, f \circ g)}{T(r, f)} \geq \frac{\frac{1}{3} \log M \left(\frac{r}{2}, f \circ g\right)}{\log M(r, f)} \geq \frac{\frac{1}{3} \log M \left(\frac{r}{4}, g\right) + o(1)}{\log r} \to \infty
$$

$$
\text{as } r \to \infty .
$$

and obtain the desired result.

To prove Eq. (11) we need one of Clunies' earlier results. If f and g are entire, then,

$$
M(r, f \circ g) \geq M((1 - o(1))M(r, g), f), \quad r \to \infty, \quad r \notin E ,
$$

where E is a set of r of finite measure and $o(1)$ and E are depending on g. As $f(z)$ is transcendental, the above inequality implies that, for any given positive constant k, since $M(r, f \circ g) \geq (1 - \alpha(1))^k (M(r, g))^k$ as $r \to \infty, r \notin E$,

$$
\varlimsup_{r \to \infty} \frac{\log M(r, f \circ g)}{\log M(r, g)} \geq k ,
$$

and k can be chosen to be arbitrarily large, the result follows.

To prove Eq. (12), we recall a well known fact in the Nevanlinna value-distribution theory; namely if f is meromorphic then for all complex number w outside a set of zero capacity depending on f,

$$N\left(r, \frac{1}{f-w}\right) \sim T(r,f) \quad \text{as } r \to \infty .$$

It follows from this result, that a constant α can be chosen so that $f(z) - \alpha$ has an infinite number of zeros, $\varsigma_1, \varsigma_2, \dots, \varsigma_n, \dots$ and

$$N\left(r, \frac{1}{f \circ g - \alpha}\right) \sim T(r, f \circ g) \quad \text{as } r \to \infty ,$$

$$N\left(r, \frac{1}{g - \varsigma_n}\right) \sim T(r,g) \quad \text{as } r \to \infty, \ n = 1, 2, \dots .$$

It follows that, for any given positive integer n,

$$N\left(r, \frac{1}{f \circ g - \alpha}\right) \geq \sum_{i=1}^{n} N\left(r, \frac{1}{g - \varsigma_i}\right) .$$

and so

$$\lim_{r \to \infty} \frac{T(r, f \circ g)}{T(r,g)} \geq n .$$

As n can be chosen to be arbitrarily large the result in Eq. (12) follows.

Theorem A.5. (Clunie) Let $f(z)$ be meromorphic and $g(z)$ be entire and suppose that $f(z)$ and $g(z)$ are transcendental. Then

$$\overline{\lim_{r \to \infty}} \, \frac{T(r, f \circ g)}{T(r,f)} = \infty .$$

Theorem A.6. (Clunie) (i) Let $f(z)$ be transcendental meromorphic and g be transcendental entire. Suppose that at least one of them is of finite order. Then

$$\lim_{r \to \infty} \frac{T(r, f \circ g)}{T(r,f)} = \infty .$$

(ii) Let f and g be given as in (i). Suppose that $g(z)$ is of finite order. Then

$$\lim_{r \to \infty} \frac{\log M(r, f \circ g)}{\log M(r,g)} = \infty .$$

Remarks. (1) By constructing an example Clunie showed that the finiteness of the order of g is a necessary condition for the validity of (ii) of Theorem A.6.

(2) Clunie also demonstrated (by example) that for a certain pair of functions f (meromorphic) and g (entire).

$$\lim_{r \to \infty} \frac{T(r, f \circ g)}{T(r, f)} = 0 .$$

(3) Adopting Clunie's reasoning, Song and Yang showed
   (i) there exists an entire function g such that

$$\lim_{r \to \infty} \frac{\log \log M(r, e^g)}{\log M(r, g)} = 0 ,$$

and
   (ii) there exists a meromorphic function f and entire function g such that

$$\lim_{r \to \infty} \frac{\log T(r, f \circ g)}{\log T(r, g)} = 0 .$$

3. THE EXTENSION OF PÓLYA'S THEOREM TO MEROMORPHIC FUNCTIONS

Theorem A.7. (Edrei and Fuchs) Let $f(z)$ be a meromorphic function that is not of zero order and g be a transcendental entire function. Then $f(g)$ is of infinite order.

The above is an extension of Polya's theorem and is an immediate consequence of the following result.

Theorem A.8. (Edrei and Fuchs) Let f and g be entire functions. Assume that the zeros of f have a positive exponent of convergence and that g is transcendental. Then the zeros of $f(g)$ do not have a finite exponent of convergence.

4. SOME NECESSARY CONDITIONS FOR THE EXISTENCE OF MEROMORPHIC SOLUTIONS OF CERTAIN DIFFERENTIAL EQUATIONS

Theorem A.9. (Steinmetz, Gackstatter and Laine) Let

$$P(z, w, w', \dots, w^{(n)}) \equiv \sum_{\lambda \in I} \alpha_i(z) w^{i_0} (w')^{i_1} \dots (w^{(n)})^{i_n}$$

be a differential polynomial in $w(z)$ with the coefficients $\alpha_j(z)$ being nonzero meromorphic functions, where I is a finite set of multi-indices $\lambda = (i_0, i_1, \ldots, i_n)$ $(i_0, i_1, \ldots, i_n$ are nonnegative integers). Let

$$A(z, w) = \sum_{j=0}^{p} a_j(z) w(z)^j, \quad B(z, w) = \sum_{k=0}^{q} b_k(z) w(z)^k ,$$

where $a_j(z)$ and $b_k(z)$ are nonzero meromorphic functions with $a_p(z) b_q(z) \not\equiv 0$.

Consider the differential equation

$$P(z, w, \ldots, w^{(n)}) = A(z, w)/B(z, w)$$

and set

$$\Delta = \max_{\lambda \in I}(i_0 + 2i_1 + \ldots + (n+1)i_n) ,$$
$$d = \max_{\lambda \in I}(i_0 + i_1 + \ldots + i_n) ,$$

and

$$\Delta_0 = \max_{i \in I}(i_1 + 2i_2 + \ldots + ni_n) .$$

If the above differential equation has a meromorphic solution $w(z)$ satisfying

$$T(r, c(z)) = \circ T(r, w) \quad \text{as} \quad r \to \infty ,$$

outside a set of finite measure E; where $c(z)$ represents any of the coefficients (i.e., c_j, a_i, b_k) in the equation, then

(i) $q = 0$ and $p \leq \Delta$ and

(ii) $p \leq d$ under the additional condition that $N(r, w) = \circ T(r, w)$ as $r \to \infty$ outside a set of finite measure.

5. SOME PROPERTIES OF DIFFERENTIAL POLYNOMIALS

Let M define the class of all the meromorphic function. We shall represent as $S(r, f)$ any quantity satisfying $S(r, f) = \circ\{T(r, f)\}$ as $r \to \infty$, possibly outside 'a set of r' of finite measure.

Let $P(z, f)$ be a polynomial in f and its derivatives with the coefficient $a(z)$ satisfying $T(r, a(z)) = S(r, f)$. We shall call $P(z, f))$ a differential polynomial in f (or simply a differential polynomial $P(f)$) and $P_n(z, f)$ denotes differential polynomial of degree at most n in f.

Theorem A.10. (Clunie) Let f be a transcendental meromorphic functions. Suppose that

$$f(z)^n P(z, w) = Q_m(z, w) ,$$

where $P(z, f)$ and $Q_m(z, f)$ are both differential polynomials in f with $m \leq n$. Then

$$m(r, P(z, f)) = S(r, f) .$$

The above result and the one below are both contained in Hayman's book *Meromorphic Functions*.

Theorem A.11. (Tumura and Clunie) Let $f(z)$ be a non-constant meromorphic function.

Suppose that

$$g(z) = f(z)^n + P_{n-1}(f) ,$$

and that

$$N(r, f) + N\left(r, \frac{1}{g}\right) = S(r, f) .$$

Then $g(z) = h(z)^n, h(z) = f(z) + \frac{1}{n}a(z)$, and $a(z)$ is obtained by equating $h(z)^{n-1}(z)a(z)$ with the terms of degree $n-1$ in $P_{n-1}(f)$ after substituting $h(z)$ for $f(z), h'(z)$ for $f'(z)$, etc.

For example, if $P_{n-1}(f) = a_0(z)f'(z)f(z)^{n-2} + P_{n-2}(f)$, then

$$h^{n-1}a(z) = a_0(z)h'h^{n-2} ,$$

and hence

$$a(z) = a_0(z)\frac{h'}{h} = \frac{a_0(z)}{n}\frac{g'}{g} .$$

Therefore, in this case

$$g(z) = h^n(z) = \left(f(z) + \frac{a_0(z)}{n}\frac{g'(z)}{g(z)}\right)^n .$$

6. A SIMPLER PROOF OF STEINMETZ'S THEOREM

The following is a simpler proof of Steinmetz's Theorem due to Gross-Osgood. The method is motivated by some techniques employed in the study of transcendental number theory.

Theorem A.12. (Steinmetz) Suppose g is entire, $n \geq 2$ is a natural number, and $f_i(z) \not\equiv 0 (1 \leq i \leq n)$ and $h_i(z) \not\equiv 0$ are meromorphic. Suppose that $\sum_{i=1}^{n} T(r, h_i) = O(1)T(r, g)$. If $\sum_{i=1}^{n} f_i(g)h_i(z) \equiv 0$, then there exist n polynomials, $P_i(z)$, not all zero such that $\sum_{i=1}^{n} P_i(g)h_i(z) = 0$.

Lemma. Let $F_1 \not\equiv 0, F_2 \not\equiv 0, \ldots, F_m \not\equiv 0$ be m formal power series in $z - a$ for any complex number a. Then there exists an infinite sequence of $(m+1)$-tuples of polynomials in z, $(Q_j, P_{1j}(z), P_{2j}(z), \ldots, P_{mj}(z))$ that satisfy, for each j, the following three properties:

(i) $Q_j(z) \not\equiv 0$

(ii) $\max\{\deg Q_j, \deg P_{ij}, \ldots, \deg P_{mj}\} \leq mj$, and

(iii) $z = a$ is a zero of multiplicity at least $(m+1)j$ for every $Q_j(z)F_i(z) - P_{ij}(z), 1 \leq i \leq m$.

Proof of the lemma. Property (iii) actually imposes $m(m+1)$ linear homogeneous conditions on the (yet to be determined) coefficients of $Q_j(z)$ and the $P_{ij}(z)$. By (ii) there are no more than $(m+1)(mj+1)$ such coefficients to be determined. Since $(m+1)(mj+1) > m(m+1)j$, it follows from the theory of system of linear equations that for each j, there exists a set of coefficients for the $Q_j(z)$ and the $P_{ij}(z)$ that are not all identically zero, such that (iii) holds. Next we show that $Q_j(z) \not\equiv 0$. Otherwise, we conclude from $P_{ij}(z) = Q_j(z)F_i(z) - P_{ij}(z)$ and (iii) that each $P_{ij}(z)$ would vanish at $z = a$ to an order greater than $\deg P_{ij}$, which would yield $P_{ij} \equiv (z), 1 \leq i \leq m$ and $Q_j(z) \equiv 0$, a contradiction.

Proof of the theorem. Set $m = n - 1$ in the lemma and $F_i(z) = \frac{f_{i+1}(z)}{f_i(z)}, 1 \leq i \leq m$. Let a be any point such that each F_i can be expanded into a power series. Define $G_j(z) \equiv Q_j(g)h_1(z) + \sum_{i=1}^{n-1} P_{ij}(g)h_{i+1}(z), 1 \leq j \leq \infty$. We are going to show that at most a finite number of the $G_j(z)$ are nonzero functions. In what follows we may assume each $G_j(z) \not\equiv 0$. This will lead to a contradiction. First we will show that

$$H_j(z) \equiv \frac{G_j(z)}{|g(z) - a|^{nj}} = \frac{Q_j(g)h_1(z) + \sum_{i=1}^{n-1} P_{ij}(g)h_{i+1}(z)}{|g(z) - a|^{nj}}$$

has exactly the same poles as does $G_j(z)$.

From hypothesis

$$\sum_{i=1}^{n} f_i(g)h_i(z) \equiv 0 , \qquad (13)$$

we have, by multiplying (13) by $Q_j(g)/f_1(g)$,

$$H_j(z) = - [g(z) - a]^{-nj} \left[Q_j(g)h_1(z) + \sum_{i=1}^{n-1} Q_j(g)\frac{f_{i+1}(g)}{f_1(g)} h_{i+1}(z) \right.$$
$$\left. -(Q_j(g)h_1(z) + \sum_{i=1}^{n-1} P_{ij}(g)h_{i+1}(z)) \right] .$$

Thus

$$H_j(z) = \sum_{i=1}^{n-1} (Q_j(g)F_i(g) - P_{ij}(g))(g(z) - a)^{-nj}h_{i+1}(z) . \qquad (14)$$

By (iii) of the lemma, for each j,

$$(Q_j(g)F_i(g) - P_{ij}(g))(g(z) - a)^{-nj}$$

is entire, so the division of $G_j(z)$ by $[g(z) - a]^{nj}$ yields no new poles (since $g(z) - a$ is entire, the division cannot remove any pole). Hence

$$N(r, H_j(z)) = N(r, G_j(z)) \le \sum_{i=1}^{n} N(r, h_i) \qquad (15)$$

Let

$$T\{z : \big||g(z) - a| \le 1\} .$$

Thus for all z in T, we can see that

$$|(Q_j(g)F_i(g) - P_{ij}(g))(g(z) - a)^{-nj}|$$

is bounded. Hence, by virtue of (14), we have, for all $z \in T$,

$$|H_j(z)| = O\left(\sum_{i=1}^{n-1} |h_{i+1}(z)|\right) \ .$$

On the other hand, by property (iii) of the lemma, for all $z \in \mathbb{C} \backslash T$. ($\mathbb{C}$ denotes the complex plane), we have for each i and j

$$\left|\frac{Q_j(g)}{(g(z) - a)^{nj}}\right| \quad \text{and} \quad \left|\frac{P_{ij}(g)}{(g(z) - a)^{nj}}\right|$$

are bounded. Hence,

$$|H_j(z)| = O\left(\sum_{i=1}^{n} |h_i(z)|\right) \ .$$

Thus, we have

$$m(r, H_j(z)) \le \sum_{i=1}^{n} m(r, h_i) + k \tag{16}$$

where k is a positive constant independent of r.

It follows from (14) and (15) that

$$T(r, H_j(z)) \le \sum_{i=1}^{n} T(r, h_i) + k = O(1)T(r, g) + k \ . \tag{17}$$

Next we proceed to estimate $m\left(r, \frac{1}{H_j}\right)$ which by Nevanlinna's first fundamental theorem will be no larger than the right hand side of (17) (possibly for a new constant replacing k). We denote this bound by B_j. If j is sufficiently large, we shall derive a lower bound for $m\left(r, \frac{1}{H_j}\right)$ which will exceed B_j and the theorem will then be proved.

For all $z \in \mathbb{C} \backslash T$, we have from (ii) of the lemma that

$$|H_j(z)| \le O(1)(|g(z) - a|)^{-j} \max\{|h_i(z)|; \ 1 \le i \le m\} \ .$$

Since the $\max\limits_{1 \le i \le m} \deg\{Q_j, P_{ij}\} \le (n-1)j$. Thus, for some positive constant d, independent of z,

$$\log^+\left|\frac{1}{H_j(z)}\right| \ge \log\left|\frac{1}{H_j(z)}\right| \ge j \log|g(z) - a| \ .$$

$$-\sum_{i=1}^{n} \log^+ |h_i(z)| - d = j \log^+ |g(z) - a| - \sum_{i=1}^{n} \log^+ |h_i(z)| - d \ . \tag{18}$$

(Note that (18) holds for all $z \in T$ as well.)

We obtain by averaging (18) over the circle: $|z| = r$,

$$m\left(r, \frac{1}{H_j(z)}\right) \geq jT(r, g) - \sum_{i=1}^{n} T(r, h_i) - d \leq B_j .$$

The above inequality is impossible to hold for sufficiently large j. This also completes the proof of the theorem

REFERENCES

1. I.N. Baker and F. Gross, "Further results on factorization of entire functions", *Proc. Symposia Pure Math. Amer. Math. Soc.*, Providence, R.I. II, (1968) 30-35.
2. J. Clunie, "The composition of entire and meromorphic functions", *McIntyre Memorial Volume*, Ohio Univ. Press (1970).
3. A. Edrei and W.H.J. Fuchs, "Sur les valeurs déficientes et les valeurs asymptotiques des fonctions méromorphes", *Comment. Math. Helv.* **33** (1959) 258-295.
4. A. Edrei and W.H.J. Fuchs, "On the zeros of $f(g(z))$ where f and g are entire functions", *J. Analyse Math.* **12** (1964) 243.
5. R. Goldstein, "On factorization of certain entire functions", *J. London Math. Soc.*, (2) (1970) 221-224.
6. R. Goldstein, "On factorization of certain entire functions, II", *Proc. London Math. Soc.* **22** (1971) 483-506.
7. F. Gross and C.C. Yang, "Further results on prime entire functions", *Trans. Amer. Math. Soc.* **142** (1974) 347-355.
8. F. Gross, *Factorization of Meromorphic Functions*, U.S. Government printing office, Washington, D.C. (1972).
9. W.K. Hayman, *Meromorphic Functions*, Oxford Univ. Press, Oxford (1964).
10. M. Ozawa, "On prime entire functions, I and II", *Kodai Math. Sem. Rep.* **22** (1975) 301–308, 309–312.
11. M. Ozawa, "Sufficient conditions for an entire function to be pseudo-prime", *Kodai Math. Sem. Rep.* **27** (1976) 373–378.
12. M. Ozawa, "On uniquely factorizable meromorphic functions", *Kodai Math. J.* **1** (1978) 339–353.

13. G.S. Prokopovich, "On superposition of some entire functions", *Ukrain. Mat. Zh.* **26**, No. 2, March-April (1974) 188-195.

14. G.S. Prokopovich, "On pseudo-simplicity of some meromorphic functions", *Ukrain. Mat. Zh.* **27**, No. 2, March-April (1975) 261-273.

15. G.S. Prokopovich, "Fix-points of meromorphic functions", *Ukrain. Mat. Zh.* **25**, No. 2 (1972) 248-260 (English translation 198-208).

16. J.F. Ritt, "Prime and composition polynomials", *Trans. Amer. Math. Soc.* **23** (1922).

17. P.C. Rosenbloom, "The fix-points of entire functions", *Medd. Lunds Univ. Mat. Sem., Suppl. Bd. M. Riesz* (1952) 186-192.

18. H. Selberg, "Algebroid functions and inverse functions of abelian integrals", *Arhandlinger utgittav det norske Videnskaps-Akademi i Oslo I. Matem.-Naturvid.* **8** (1934) 1-72.

19. G.D. Song and C.C. Yang, "On pseudo-primality of the combination of meromorphic functions satisfying linear differential equations, in value distribution theory and its applications", edited by C.C. Yang, *Contemporary Math-series* **25**, American Math. Soc. Providence, R.I. (1980).

20. N. Steinmetz, "Über die fakorisierbaren Lösungen gewohnlichen Differentialgleichungen", *Math. Zeit.* **170** (1980) 169-180.

21. N. Toda, "On the growth of meromorphic solutions of an algebraic differential equations", *Proc. Japan Acad.* **60**, Ser. A (1984) 117-120.

22. H. Urabe, "Uniqueness of the factorization under composition of certain entire functions", *J. of Math. of Kyoto University* **18**, No. 1 (1978).

23. H. Wittich, *Neuere Untersuchungen über Eindeutige Analytische Funktionen*, Springer-Verlag, New York (1984).

24. H. Wittich, "Ganze transendente Lösungen algebraischen differentialgleichungen", *Math. Ann.* **122** (1950).

25. C. Yang, *Factorization Theory of Meromorphic Functions*, Lecture Notes in Pure and Applied Mathematics, Vol. **78** (edited by C. Yang), Marcel Dekker, Inc (1983).

INDEX